TALKING TO ANIMALS

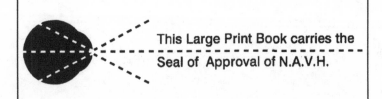

This Large Print Book carries the
Seal of Approval of N.A.V.H.

TALKING TO ANIMALS

HOW YOU CAN UNDERSTAND ANIMALS AND THEY CAN UNDERSTAND YOU

JON KATZ

THORNDIKE PRESS
A part of Gale, Cengage Learning

GALE
CENGAGE Learning·

Farmington Hills, Mich • San Francisco • New York • Waterville, Maine
Meriden, Conn • Mason, Ohio • Chicago

GALE
CENGAGE Learning®

Copyright © 2017 by Jon Katz.
Thorndike Press, a part of Gale, Cengage Learning.

LIBRARY OF CONGRESS CATALOGING-IN-PUBLICATION DATA

Names: Katz, Jon.
Title: Talking to animals : how you can understand animals and they can understand you / by Jon Katz.
Description: Large print edition. | Waterville, Maine : Thorndike Press, a part of Gale, Cengage Learning, [2017] | Series: Thorndike Press large print popular and narrative nonfiction
Identifiers: LCCN 2017012254 | ISBN 9781432838270 (hardcover) | ISBN 143283827X (hardcover)
Subjects: LCSH: Human-animal communication. | Large type books.
Classification: LCC QL776 .K38 2017b | DDC 591.59—dc23
LC record available at https://lccn.loc.gov/2017012254

Published in 2017 by arrangement with Atria Books, an imprint of Simon & Schuster, Inc.

Printed in Mexico
1 2 3 4 5 6 7 21 20 19 18 17

To Maria and the Animals

CONTENTS

INTRODUCTION

I was walking on a path in the deep woods by the big old Civil War–era farm I had impulsively purchased a few months before. It was just after dusk, and the light was fading. Enchanted by the dark, soft feel of the forest, I entered a beautiful and alien world. I was so glad to be there, and even happier that my new dog, Rose, was coming with me.

Rose was a young border collie who had just come to me from Texas. I had great hopes for her as my farm dog. In just a few days, Rose had grasped the concept of taking a walk. When I left the house, she ran eagerly alongside me and stayed close, as great working dogs will instinctively do.

One day, we were out walking, enjoying our time together, when suddenly the atmosphere on the path changed, grew tense. Something was happening, something I could sense but not readily see or hear in the gathering darkness. Rose had stopped. A busy, curious, fast-moving dog, she was frozen in place. She had gone into a crouch, but it was not a herding crouch, which is low and tense and close to the ground. Her hair was standing up on her neck; so, I realized, was mine.

Rose had moved directly in front of me, as if to protect me. I didn't understand what was going on until I looked beyond and saw three pairs of gleaming yellow eyes, staring right at us, a hundred feet straight ahead.

Three coyotes, right there on the path, as if they were expecting us. I was not prepared for such an encounter. I had never seen a coyote, but I'd been told that coyotes avoided humans, and never confronted them. I expected them to run at the sight of this big man and his dog. They did not. They looked calm, curious, determined. So did Rose. The dog and the coyotes mirrored one another, their postures almost identical.

My biggest fear was that Rose would charge ahead, choose to fight, and try to fend the coyotes off as if they were sheep. I had already seen how protective Rose was of me. Once, when a ram charged me from behind (they do that sometimes, since they are full of testosterone) and knocked me into a fence post, this little dog had dug a hole under the fence that was penning her in, charged up the hill, grabbed on to the ram's testicles, and run the poor screaming creature down the hill.

Rose did not back down from anything.

I had no idea what to do. I thought of breaking off a branch or picking up a stick, but worried that any movement might set off an attack. Rose stepped forward a few feet. I told her to "stay" in a loud, clear voice. It was a request, not a command. The forest had gone deathly quiet.

Instinctively, I froze and decided to be patient and still. I took a deep breath, cleared my head. I knew not to turn and run, but I was not sure I could control this young and strong-willed dog. I was still just a city boy, and I was not prepared for this.

Then, a curious thing happened.

I closed my eyes and cleared my head. *Forget everything you have known or thought about the woods, dogs, coyotes, fear,* I told myself. *Take a deep breath, think strength, feel strong.* I had no commands to give Rose, no words, but I had images and I painted a sketch in my head of what I wanted to happen, what I was sure would happen. Project confidence, I told myself. Be sure, be clear. I imagined Rose still, ears up, tail up, back straight. Border collies

11

know how to stare. I had already seen her intimidate sheep. *Stare. Don't scream at her,* I told myself. *Don't shout words she doesn't comprehend, spew commands she doesn't know.*

In my head, there was this scene: Rose staying absolutely still, meeting the coyotes' gaze, showing strength, but not aggressiveness. She would not run, she would not fight. She would communicate that it was her path, her human. That she had a right to be there and would stand her ground. The coyotes, clever opportunists, could go find some easier and safer target for their food.

Right then, she did exactly what I imagined her doing. She stood her ground, held her gaze and posture. She looked absolutely confident and resolute, as if this were her path and there was no way she was giving ground. She could have done a hundred things — run, barked, fought, growled — but she did precisely what I wished her to do.

Another set of images came into my head as we stood there. These came from Rose's point of view, from low down on the ground, level with the eyes of the coyotes. The images I was seeing in my head were curious, disorienting, not really aggressive. I painted a mental picture of what I wanted to happen. I used all of my senses to imagine it. I cleared my head of everything but the present. I imagined stillness, calm. I thought of strength, determination, absolute confidence. I thought this again and again, until it became a feeling, and I sent it to Rose: *We are okay, we are leaving, we are going home.* I pictured the path clear, the coyotes gone.

Rose growled, then whined, but she did not move. She looked absolutely peaceful and confident, not so much aggressive as at home, in command. She was barely breathing, still as a stone.

Still, the yellow eyes seemed to shine all the more brightly, the coyotes all the more fearsome. Despite my attempts to radiate calm, I was soaked with sweat, terrified of what might happen if Rose bolted and ran ahead, as she had seemed poised to do.

Then, in a blur, the scene changed.

The coyotes were gone. They had simply vanished. I never saw them move, or where they went. But I would not soon forget those yellow eyes.

Rose visibly relaxed, but was vigilant. She watched the path, listened to the woods, smelled its stories. I could see that she was also gathering the story of the coyotes, a story that was beyond me and my inadequate instincts. She moved forward slowly, sniffed the air, and then we both turned around and returned quickly to the farm, each of us pausing every few feet to turn around and look back.

It was only later, back at my new farm, sitting on my porch, that I realized that something powerful had occurred. Rose and I had spoken to one another on that path. We had begun what turned out to be a great conversation that was to change both of our lives, that would save my life more than once, and help me survive, even thrive, on a ninety-acre farm with sheep, cows, goats, dogs, and other animals, through joy and travail, blizzards and dramas, escapes and terrors.

My conversation with Rose was to grow beyond my imagination, transform my ideas about animals, and open my eyes to the possibilities of understanding them and of learning how to truly communicate with them, to forge a connection that transcends the need for training, pleading, shouting, and manipulation.

Rose and I had begun to understand one another in a completely unexpected way, one that evolved from a singular primal exchange on that path to a great, complex, diverse, and wondrous dialogue that I have been able to cultivate with so many different animals over the last several years.

There was Orson, another border collie. Elvis, the three-thousand-pound steer. Minnie, Mother, and Flo, the barn cats. Rocky, a blind pony. A love dog named Lenore. A hospice dog named Izzy. A brave rooster named Winston. Numerous chickens; goats that often seemed smarter than me. A rescue donkey named Simon. Two female donkeys, Lulu and Fanny. An abandoned hell dog from the Adirondacks named Frieda. A struggling mother sheep named Ma. Lambs named Liam, Deb, and Jake. There was also Rose's true successor, Red, a border collie from County Tyrone, Northern Ireland. And, finally, Fate, the puppy who represents the culmination of my long and fascinating work with animals, including the way animals can communicate with one another and with us.

Each of these animals has taught me something. Sometimes it is about listening, sometimes about talking. Often it is about me.

What you will find in the following pages is not an animal training book. It is not a how-to book or a science-heavy study of animal behavior. It is not a book about obedience or how to get your dog to do exactly what you command. It is also not a book about animal rights. I am as uncomfortable with animal politics as I am with people politics. I do not argue my beliefs on Facebook or anywhere online. I am not here to mind your business or tell you what to do. Instruction is not the purpose.

My hope is, simply, that readers of this book will gain an understanding about how to live well with the animals they love, and understand them in a better way. The things I have learned over the last several years are not magic or voodoo. They are not impossible to replicate. Anyone can become more connected to the animals in his or her life.

There are enormous economic, emotional, and environmental benefits to understanding our animals and supporting them. We can't help them and keep them in our world if we can't communicate with them. But in order to do that, we need to know them as they really are, not as the emotional fantasies we construct about them.

This is a very personal book, a spiritual book, an anecdotal account of my own journey over many years to a wiser and more mystical understanding of animals. I am not a guru or an all-knowing animal vizier; I don't believe any such thing exists. I can only relate my own rich experiences and hope they are of some use and benefit to those who are interested in connecting with animals in a deeper way.

Training is a spiritual experience, not an exercise in domination. But of course nothing is black-and-white. Dogs live in a world that can be hostile and dangerous for them. There are times when they simply need to do what they are told, and quickly, for their own safety and the safety of others.

But more important, training is a way of knowing animals, loving them, and helping them to live safely in our world. There are several tools that can be used to communicate effectively with animals, among them food, body language, attitude, and visualization.

Early on, I learned never to underestimate the importance of food as the gateway to communicating with animals. It is the foundation of trust and attention. If animals don't trust us, they will never pay attention to us; if they cannot pay attention to us, they cannot listen. Food, thoughtfully and judiciously applied, builds trust and attention.

I first learned the importance of body language with Elvis, my three-thousand-pound steer, a monstrously huge animal. Steers are not known for talking or listening, and farmer after farmer warned me to stay away from him. These animals are untrainable, I was told.

But Elvis was not untrainable. I learned how to gain his trust with apples and oranges and how to move in a way that brought him to a full stop from a fast trot without plowing into me. He came when I called him, stayed when I told him, even stood still for the application of ointments and antibiotics for the scratches and cuts he would accumulate while grazing.

Food is necessary to establish trust, but attitude is perhaps the most important tool of all. Animals read intentions. If you mean it, they know it; if you don't, they sense it. It is essential to have your intentions clear in your head, and in the attitude you project to the animal.

While visualization may seem the most abstract of all the communication tools, and the idea of "image-talking" far-fetched, it is actually heavily grounded in research and behavioral science dating all the way back to Aristotle. For some years now, behaviorists and researchers like Temple Grandin (*Animals in Translation*) have been exploring the link between the way dogs think and the way autistic children think — namely, they both think and communicate in images.

People throw a lot of words at overwhelmed animals, but animals do not communicate in words. According to biologists and behaviorists like Grandin, they communicate in images, and the images that course rapidly through their consciousness trigger their powerful instincts. Frank Niles, a well-known social scientist and adventure athlete, says that visualization in its simplest form is simply a technique for creating a mental image of a future event. It is the art of imagining what we want. When we visualize our desired outcome, we begin to believe in the possibility of achieving it. Studies show that visualization can also help reduce anxiety and fear.

Animals are extraordinarily sensitive to our feelings and movements and smells and emotions. Through all of these senses they can grasp this determination, these images that we are trying to transmit to them. There is nothing woo-woo or psychic about it. It does, however, require awareness, concentration, and discipline.

It is difficult for a human like me to think in images, even more so as a writer deeply invested in the authority and power of words. But they are largely useless when it comes to communicating with animals. Opening up to the idea of image-talking can be a transformative experience. I have used this successfully for years. But I'm still learning every day.

Not too long ago, most Americans lived with animals and knew them well. Now the vast majority of Americans live on the coasts, disconnected from the natural world and the real lives of animals. Over time, our view of animals has been emotionalized. Increasingly, we view them through the prism of dependence, of rescue and abuse. We seem to need to see them as piteous, helpless, and dependent; this intensifies our connection to them and offers us many needed ways to feel loved, connected, and worthwhile. But it does not always foster understanding.

At age fifty, after spending several years as a journalist in Washington, D.C., on a hiatus from engaging with the creatures who meant so much to me in my adolescence, I resolved to try to bridge that gap in my life. I moved to the country to reconnect with animals and the natural world. Every day since, I have learned something new about how to listen to animals and talk to them in ways that benefit us both.

My animals are healthy, content, affection-ate, and responsive. They do not harm people or other animals. Our lives together are a symphony of compromise and under-standing.

For about a decade now, I have used these techniques — food, body language, attitude, and visualization — to teach my dogs to stay away from the road. To never cross a road without permission, to never wander near it, to never chase a dog, ball, toy, rabbit, or squirrel if it means crossing a road.

I have applied this approach to eight dogs as of this writing: three Labs, four border collies, and a Rottweiler-shepherd mix. Each of these dogs had a tendency toward explosive chasing — balls, sticks, animals, other dogs, me.

Lenore, my black Lab, was a boisterous and enthusiastic dog who will run through barbed wire to retrieve a ball, and who loves to blast out the door each morning when I let her out of the house. When Lenore was eight months old, I started taking her out by the farmhouse and toward the steep road that runs right by the house. This is an especially dangerous road, because cars coming downhill pick up speed, and often have built up too much momentum to stop easily or quickly.

I bought a bag of beef jerky and each morning dropped some at a point about fifteen feet from the road. After the third or fourth day, Lenore would bound outside, find the beef jerky, and stop there, hoping for more. If she waited there for me, I would toss her some more, and that quickly became our habit. Soon she was much more interested in the jerky than the road.

After about ten days of this, I changed our procedure. I would drop some jerky at our spot, then I would draw my hand in front of her and say "stay." If she stayed for three minutes or more, she got more beef jerky.

That was the food part; the basis of effective communication, the building of trust and attention. Once she grasped the idea of staying there, it happened automatically without the beef jerky. We stayed for at least three minutes every time we came out. I would position myself between Lenore and the road, hold up my hand, stand tall and straight, and say "stay" with authority and conviction. Sometimes I would reward her with a treat, sometimes not. She never knew for sure if she was getting one.

My body language was clear. I was stiff and present, right between her and the road. I held my hands wide, a position most animals respect. If she inched a bit toward the road, I would stamp my foot and glower. That was more of a setback for me than it was for Lenore. My impatience was a sign that I needed to get my head straight. I would tell myself, *we can do this, I can make this happen, I will succeed.* She would not go near the road, not ever, not for a ball, a chipmunk, a piece of steak. And I meant it.

Then I did some image-talking. I imagined Lenore stopping in this same spot every time we came outside and sitting there waiting for instructions. I had paved the way with food, solidifying my efforts with positive reinforcement, with body language, and the "stay" command. Now I was adding the final layer of visualization: I cleared my head and focused on what I wanted to happen.

Dogs are Darwinian marvels — they have adapted and thrived for thousands of years by figuring out what humans want and need and by doing it. They want to please us; it is how they survive. Our task is to make them understand what we want and need. And in return, they can live safely with us and we will take care of them, feed them, and provide them shelter and attention and affection.

If we do not understand animals, we cannot really know how to help them live safely and remain in our world. In 2014, the World Wildlife Fund reported that half of the animals on the earth have vanished since 1970. We have few viable ideas or proposals for helping animals because we often do not understand any longer what it is they really need.

In recent years, I have been writing about one of the most significant conflicts in the animal world, the movement to ban the carriage horses from cities like New York due to a widespread belief that work is cruel for domesticated animals like the horses, is even abusive.

But the proponents of these bans do not seem to know that these horses have never lived in the wild, have been bred to work, need to work to be active and healthy, and adapt better than any other domesticated animal — even dogs — to urban life. What they do need is work and stimulation and shelter and healthy food and good medical care. And what people need, children especially, is more exposure, not less, to these magnificent creatures.

But so many people do not know that because rather than communicate with the horses and learn to understand their needs, we project our emotions onto them and rush to rescue them, sometimes destroying them and their lives in the process. So there are good reasons for talking to animals. Their very lives depend on it.

Since I've worked with her, Lenore has never wandered past her "spot," she has never run into the road, she has never chased anything into the road. I have even tested her. I've put food across the road, tossed a ball across the road, run across the road myself and tried to get her to chase me. She has never once run across the road without an explicit one-word release command — "cross." The point of single-word commands is that they are quick and simple for the dogs to understand amid all of our human chatter. That's why herding dogs like border collies often have single-syllable names.

Careful use of language is critical in learning how to communicate with animals. We humans tend to use too many words, many of which echo our frustration, anger, and confusion. It is important to use as few words as possible, and to associate them with the behaviors we need and seek.

There are so many benefits to learning how to communicate with animals. Love, trust, a spiritual connection that goes to the heart of the human–animal bond. A connection to nature, to the healing and nurturing element of life with animals. Every time I listen to them, I learn about myself. I evolve and grow.

This was the case with Red, a dog who had never lived in a house until he came to my farm. He was not housebroken; he had always lived outdoors. Linoleum floors terrified him. Red and I were perfectly suited for the great experiment we were about to undertake together, an epic journey into the heart of animal–human communication. He was a remarkably intelligent and adaptive dog, curious, eager to please, used to solving problems. I am an impulsive and impatient man, easily distracted. But Red and I fit together like two pieces of a puzzle.

When he arrived on our farm, Red had to acclimate himself to living alongside three donkeys — guard animals who thought he was a coyote stalking the sheep — and Rocky, a blind Appaloosa pony who had lived alone in a field for fifteen years before joining our farm. Red had to learn to stay off the road, ride in a car, and become comfortable among people.

He also had to learn the difficult and exquisitely precise protocols of animal-based therapy work. At our farm, we work with veterans of the wars in Afghanistan and Iraq, and when we are in session, he and I have to communicate in the clearest and most reliable of ways.

We once saw a young veteran with severe post-traumatic stress; even the smallest noise or movement could traumatize him. Red approached him, and I needed Red to be very slow and deliberate in his movements. So I held my hand up to signal him to pause, and waved gently to get him to move forward. I said "sssssssh!" in a soft voice so he'd know to be quiet.

I didn't have any commands for this situation — it depended entirely on our ability to communicate and our understanding of one another. Red approached the young man slowly, watching him, but also looking at me, listening to me. When he got close, he put his head on the young man's knee and waited for him to respond. That way, the soldier had time to prepare for him. By the time Red approached him, the soldier was smiling, ready. I had not spoken a word or given a single command.

It was a beautiful connection, and a successful therapy session.

Red continues to influence my ideas about the human–animal relationship. Recently, I took him out for one of our daily pre-writing walks, this time in a beautifully tended local cemetery with tall trees and winding trails.

Red has never been on a leash with me. He walks easily alongside me, moving as I move, keeping an eye on where I am. He ignores other dogs and does not run off after strange smells and sounds. It is a joy to walk with him. We walked far this particular morning. I was tired and distracted and didn't at first notice the family of grieving mourners standing by a fresh grave a dozen yards in front of us. They were quiet, except for a young girl crying.

Red is a sensitive dog; he intuitively avoids people who do not wish to interact with him or be bothered. But I saw him stiffen at the sight of this family. His ears went up, and he turned to look at me.

I shook my head back and forth and held up my hand, as if to signal "stay." He responded to that command, and quickly. In my mind, using the image-talking that had always worked with Red before, I projected him staying with me, walking past the mourners.

I was moving off to the left when I saw him look at me again in a pleading and uncertain way, as if to get me to reconsider. I have learned over many hard years of work with my other dogs that it is important to listen carefully, to receive their messages. As

Red looked at me, an image came into my mind of Red walking up to that young girl, of comforting her. It's okay, he seemed to be telling me; she would welcome it, it would be helpful, she needs me.

It was a confusing situation for me. I had a difficult decision to make. I would never want to be responsible for a dog of mine disrupting one of the most private and sensitive of human experiences — mourning at a graveside. It was a privilege to be allowed to walk in the cemetery with a dog; I did not wish to violate that trust. Yet I was hearing something from Red about his instinct, about human need.

Because of their powerful instincts and exquisite sense of smell and sight, there are things dogs can intuit much better than humans. While humans are limited by words and visual cues, animals have so many more tools with which to read emotion. They can sense our moods a lot faster than we can sense theirs.

Over time, I have learned to trust Red. I have seen him in his therapy work; he is exquisitely sensitive.

So that day in the cemetery, I waited a minute, then dropped my head down and nodded. I pictured an image, what I wanted to happen — an image in which he went up quietly and approached the girl. I was nervous. I would be so sorry to have gotten this wrong.

Red understood me, he understood that I was giving him the freedom to make his own decision. He walked quietly, tail wagging softly, across the dirt road and toward the young girl who was standing by her family.

She looked to me like she was a teenager. She was sobbing, her head in her hands. Red went right up to her very quietly and put his head against her knee. Startled, she looked down, let out a gasp, smiled, and knelt to the ground.

"Oh, hey there," she said, and I saw a woman I took to be her mother look up and smile. Red leaned into the girl, and the two hugged for what seemed to be the longest time. I nearly cried myself, I was so moved by this tableau of love and comfort. Red knew what he was doing, knew what this young woman needed and wanted.

He had let me know, and I had listened. It was a rich experience for me. I felt so fortunate to have a dog with such assurance and such a strong talent for healing.

I walked on, and then turned and waited a few minutes. The young girl continued to hug Red. After a while, she stood up, leaned over to kiss him on the nose, mouthed "thank you" to me, and turned back to her family. Red did not need any command. He walked quickly back to me and we continued our walk.

This kind of communication has been the most compelling and powerful experience in my life with animals. We know so little about the animals we live with. We often seem at odds with them and their powerful instincts. We struggle to make them obedient, but we often fail to see how much farther we can go than that.

If there is a godfather of *Talking to Animals,* it would be the author and naturalist Henry Beston, who wrote the wonderful book *The Outermost House* nearly a century ago. Widely credited as being the inspiration for the animal rights movement, Beston called for a "wiser and more mystical understanding" of the animals left in our world. "For the animal shall not be measured by man," he wrote. "In a world older and more complete than ours, they move finished and complete, gifted with extensions of the senses we have lost or never attained, living by voices we shall never hear. . . . They are not brethren, they are not underlings; they are other nations, caught with ourselves in the net of life and time, fellow prisoners of the splendor and travail of the earth."

This is the best statement I have ever read about animals, the wisest and most inspiring. I hope to answer Beston's call.

Beston was not a political person. He did not imagine an animal rights movement as it has evolved in our time, and I doubt he would have cared for it. There was nothing angry or judgmental about him. But he was very much on my mind when I set out to learn how to really talk to the animals in my world, my partners in the joys and travails of my life.

I've worked hard almost every day for the past twenty years to look at my animals in a wiser and more mystical way. I listened to vets, and to behaviorists, trainers and breeders, rescue workers, and even animal hoarders. I spent thousands of hours working with border collies. I helped bring a dying donkey back to life, pulled lambs out of ewes' bellies in the dark of night, grappled with the energy of goats, learned how to communicate with flighty and unpredictable sheep. I took notes every day and thousands of photographs as well.

Animals, and dogs in particular, have become a complex and very emotional part of our lives. Complicated and expensive training manuals and videos are sold by the millions. Increasingly, we see animals as piteous creatures in need of rescue and intervention. We want them to have perfect lives, better often than we have.

We are projecting our words, thoughts, and emotions onto the animals, medicating them for human-style neuroses, yelling at them in frustration. Animals are personified; we see them as versions of ourselves, with our thoughts and emotions. We approach animals mostly in terms of training them to do what we say, a system of communicating that has, to my mind, failed most people and animals in the most profound way.

In many cases, standard methods of training diminish animals, treating them as spiritless, stupid, and clueless beings. This does a great disservice to them and to us.

For me, communicating with animals is a sacred challenge and responsibility. It is never about obedience; it is a spiritual and mystical thing. It asks us to change much of what we have been led to believe about how animals think.

The possibilities and rewards for people who learn to talk to their animals and listen to them are beyond the imaginations and experience of many of us.

Humans and animals have become painfully estranged from one another. There are very few animals who will not flee at the sight of a human being. We have pushed them out of our lives, and often out of our world.

There is a wiser and more mystical understanding of animals to be had, but it can only come if we can learn how to talk to them and listen to them. There are so many benefits to learning how to communicate with animals. Love, trust, a spiritual connection that goes to the heart of the human–animal bond. Every time I listen to them, I evolve and grow.

1
TALKING TO LUCKY

I always have the same dream about Lucky; I've had it on and off for nearly sixty years, since I was eight or nine years old. In the dream, Lucky is curled up in a ball in a cardboard box in the basement of the school where I first saw him. He is small, white, sweet; he chews on my finger, wags his tail. "Hey, Lucky," I say. "I'm taking you home. Talk to me."

These were the first words I ever remember speaking to an animal. I still carry this radioactive seed of memory. The image of this tiny little creature, looking up at me with hope and love, struggling to lift his head up to push against my hand, has been etched in my consciousness more than any other childhood memory. At the time I didn't know that he was responding to me, but I would come to understand the message soon enough: "Remember me," he said. My life with animals began with Lucky.

Attachment theorists would say it began some years before that, in the earliest stages of infancy, when lonely and frightened children first experience animal dreams and fantasies, and embrace the idea of animals as beloved and special friends.

But my conscious life with animals began with Lucky, when I was a miserably awkward and unhappy student at Summit Avenue Elementary School in Providence, Rhode Island.

I lived on the poor end of the east side of Providence, an Irish and Jewish immigrant neighborhood. Providence was a stern, gritty Catholic city. The Providence public school system was the gateway to education and assimilation for the children and grandchildren of immigrants, as public schools were for so many American children.

Summit Avenue School was an imposing industrial brick structure typical of urban public schools at the time. The halls were wide and shiny, filled with echoes. Boys and girls each had their own entrances and play areas. The teachers at Summit Avenue seemed old and severe to me. There was always tension between the children and grandchildren of immigrants and the children of those who were here before them. Classes were generally joyless affairs, lots of lecturing by humorless teachers and the scratching of chalk on a big green board. It was our duty to go and learn, theirs to try to ram some information into our mostly unreceptive brains.

I was lonely and strange and without a single friend in the school or outside of it. I was frightened much of the time, a bed wetter, and a physically awkward boy. I was terrified of a lot of typical adolescent activities — gym, recess, speaking up in class, getting vaccinations, doing homework, walking home alone, speaking to girls.

My family life was difficult — with my parents quarreling constantly — and I was afraid to go to school, where I was often chased and beaten up by bigger, older kids who ridiculed me and made it necessary for me to take elaborate and circuitous routes to get home safely. Many afternoons, I hid in the vast cemetery near our house. I had no friends, and was almost paralyzed by any kind of social interaction.

And then there was the abuse that is so often linked to bed wetting. Sexual and physical and emotional, it shaped so much of my childhood and my life. The point isn't what happened to me, but how I have moved past it. Lucky was an angel who came into my life to help me move forward, away from all of that darkness.

The story of Lucky and me began at school one cold gray New England morning. My classmates and I sat shivering at our desks while the ancient radiators hissed and creaked and began the long process of warming us nearly to death in our seats. It was there I learned to drowse whenever anyone gave lectures or speeches, a habit I carry still.

I was sitting at my shiny brown school desk, staring at the carved initials of countless hapless students who had come before me and doodled their initials for posterity. I was already nodding off as the interminable daily announcements began over the school loudspeakers.

I paid little attention to the morning announcements, which were followed by a mass declaration of the Pledge of Allegiance, and a scratchy record playing the national anthem. But one announcement that morning made me sit up and listen.

"Students," said Miss McCarthy, our teacher, "one of our families has a seven-week-old puppy that needs a home. The first student who arrives at the boys' entrance on Monday morning at seven a.m. can take this puppy home. Mr. Wisnewski, our jani-

38

tor, will be present." Our teacher explained later that the puppy would be at the boys' entrance because it was understood that no girl would wish to get up so early and walk to school in the dark.

It was a different world, of course. No discussions, parental notes, or permission slips were required. No one wanted to know if we had a fence, were home all day, believed in spaying or neutering, or had even consulted our parents. If you got there first, you could have the puppy and take him home, no questions asked.

I wanted this puppy more than anything; it seemed I had been waiting my whole life for him. He was mine. I had to have him.

We had once owned a German shepherd named King, but I was very young at the time and had nothing much to do with him. My father let him out in the morning, and in at night; he slept in the basement and never set foot in our house.

My parents did not spend money on dogs. King was not neutered, he was not rushed to the vet when he got sick; he holed up in the basement until he got well. There were little Kings running around all over the place. King was never walked or put on a leash, and my father would have chopped his arm off rather than walk around the neighborhood picking up poop and putting it in a plastic bag.

One day King did not come home. There were no posters put up in store windows or on telephone poles. He was responsible for himself. A neighbor told us months later that she had seen him get hit by a truck, his body hauled away in a garbage truck. King was never mentioned again.

We did not have warm and open discussions about things like dogs at the dinner table in my house. My father was not around much and paid little attention to domestic life. My mother worked, cooked, and ran the house.

I knew the decision about Lucky would be up to her, and I also knew I would be getting that puppy no matter what anybody said.

I found my mother in the kitchen after dinner — she always seemed calm and happiest alone in the kitchen doing the dishes, singing and talking to herself. I told her about Miss McCarthy's announcement.

"Absolutely not," she said. "You are too young to have a puppy, and I have enough work to do." Despite her response, I never doubted for a second that she would eventually say yes. This was just the requisite dialogue we had to get through.

She said no at least two or three more times. She sounded angry, aggrieved. Who would be responsible for the dog? Clean up after it? There was no money for vet bills. She didn't want any dog in the living room or near the furniture, or tracking up the floors or raiding the garbage cans. Who would be responsible for that?

I knew that my mother loved dogs; she was always stopping to pet them and coo at them. I knew how much she had loved King, and how sad she seemed when he was gone, even though she never spoke of it.

Back then, and for many thousands of years before, dogs lived at the periphery of life, not at the center. It is hard to even imagine a time when dogs and cats were not so intensely a part of our emotional lives. When they were kept around mainly to keep burglars away or catch mice.

America was in the midst of a great transition in the human–animal bond after World War II. Our relationship with animals was changing. The working animal was giving way to machines and cars; the wild animal was being subsumed by human development; the postwar period marked the beginning of the rise of the pet. The pet became a member of the family, and a multibillion-dollar phenomenon that has profoundly affected the way we live.

When I was a kid, dogs did not have human names and were not considered children. Animals were not family members. It would have been outrageous to suggest they were.

Dogs ate table scraps and often got hit by cars or vanished. If they got sick, they most often died, were put down, or, if one lived in the country, were taken out back and shot. There were no treats, no toys, no animal insurance plans. People got bit all the time, and female dogs had litter after litter of puppies, usually distributed free to neighbors and relatives.

My mother's dance with me went on for an hour or so, as I made one pledge after another. I'll take the dog out. I'll train him. I'll clean up, I promise. I'm sure she knew better; I know she wanted me to be happy. I saw her work her way through sputtering complaint to a softer stance.

I told my mother how much I wanted the dog, how much it would mean to me. I imagine she thought a puppy would be good for me, since she was always urging me to "step outside" of myself and join the world beyond my room, where I was invariably holed up with my books and my tropical fish.

So without exactly being agreed to, it was agreed to. She must have talked with my father about it. Nobody said no, which in that world meant yes. I could barely get through the week or sleep, I was so distracted with thoughts of my puppy. I named him Lucky because of my luck, not his. His entry into my life marked a turning point that would change the way I thought of myself.

That Sunday night, my mother loaned me her big old bell alarm clock, whose ticking kept me awake before the alarm had a chance to go off. "Good luck," she said. "Be careful crossing the streets in the dark."

Most people who love their dogs are not inclined to dwell too much on why, or how their intensity of feeling came to be. In Darwinian terms, dogs make no sense. We no longer need them for protection or help in hunting. But we love them more than ever.

People often psychoanalyze dogs, trying to determine what they are feeling and thinking. But I always found it more interesting to apply that kind of analysis to the people who own them. I write as much or more about the people who love and live with dogs as I do about the dogs themselves.

That has always been what fascinates me the most: Why do we choose the dogs we choose? Why do we love the cats that we love?

This is where attachment theory comes in. Attachment theory is important when it comes to talking to animals and listening to them. It is the first step in learning to understand and communicate with them. Attachment theory helps us understand our need and love for them, the nature of our relationship with them. It is about self-awareness, the key to living with animals in a meaningful way. It explains everything that is important about Lucky and me.

Attachment theory is the seminal study of the dynamics of long-term relationships and emotions in human beings. It is the joint work of psychologists John Bowlby and Mary Ainsworth in the 1960s, and has generally supplanted Freudian theory as the primary theory about the development of human emotions.

Bowlby revolutionized psychiatric thinking about emotions and early development, especially among preverbal children whose feelings are affected by fear, loss, or separation from their mothers. He believed that the template for most of our emotions — our security, anxiety, need for love — is shaped in the very first months and years of life, by the way in which our parents respond to our fears and loneliness.

Animal behaviorists, psychologists, and trainers have applied attachment theory to our relationships with animals. In that way, it can help explain why we attach to a particular dog or cat, why we need to rescue some dogs or hunt with others, why we love small dogs or big ones, why we only want one or have a dozen.

Attachment theory encourages us to understand the emotions and traits that we bring to the relationship. I once had a border collie named Homer, an awkward and fearful dog, or so I thought. He always seemed to lag behind, cowering at strange noises, other dogs, and loud people. I found myself yelling at him all the time, and soon I came to see I was making his problems worse. I was just not connecting with him in the way I connected with almost all of my dogs. One day — after shouting at him all during a walk to catch up, keep moving, stay with us — I stopped to ask myself why I was so angry with him.

All of a sudden, on this cool and sunny morning, it hit me that the voice I was using was not my own — it was my father's voice.

My father was a good man, but a critical man. He believed lectures would solve the complex problems of children. He sometimes considered me to be a sissy, a disappointing child, bad at sports, with few friends, a bed wetter, awkward, and inept at any kind of physical work.

When I was eleven, he threw a baseball at me during our forced "catch" sessions, and hit me in the head and knocked me down. When I came up crying, he told me I was weak and had no real strength of character. I walked off the field, and our relationship never really healed or recovered from that day. We didn't speak comfortably again for three decades.

Here was the key to what had been happening that morning with me and Homer. I was seeing Homer the same way that I was seen, as weak and fearful. A sissy. I never spoke to my daughter or any other person in that way, but here, with this poor little dog, it was coming out, the same voice, the same manner, the same anger and frustration. I realized that maybe, like me and my father, the two of us, Homer and I, just weren't meant to have a healthy relationship.

I was living in northern New Jersey at the time, and luckily, there was a young boy down the street, named Jeremy, who loved Homer. He thought he was the most wonderful dog in the world. So I gave Homer to Jeremy. With Jeremy, Homer got all of the love and affection and attention that I was not able to give him. Homer lived happily with Jeremy for twelve more years.

Some of my friends and readers were shocked that I had given one of my dogs away. It is one of those taboos that exist in parts of the animal world. But I think it was the most loving thing I had ever done with an animal, and I had John Bowlby to thank for it. If I had not been familiar with attachment theory, I would never have been able to identify the root of why I had trouble connecting with Homer. I would have condemned this sweet creature to a life of tension and frustration.

People ask me all the time how I choose a dog. Simple enough, I say. I get the dog I want, for my sake and theirs.

Our emotional interactions with dogs are mostly a replay of our own early emotional development, and we generally treat dogs and other pets in one of two ways: the way we were treated as small children, or the way we wish we had been treated.

Through the prism of attachment theory, I have since come to understand why I am drawn to certain types of dogs — border collies and Labrador retrievers, in particular. The frenetic, ADD quality of border collies, their drive to work, their curiosity, and their great loyalty are traits I value, things that I need. The Labs offer me the unconditional love I have always sought in life, and not always found. They can slip into my life and stay there.

Attachment theory asks us to look within ourselves, and through our own emotional histories, to understand our relationship with animals.

I know that every animal I have loved has challenged me to look within myself, to understand my own intuitions, instincts, strengths, and weaknesses before I can begin to understand theirs.

In 2012, I was speaking at a fund-raising dinner for an animal shelter in Palo Alto, California, and the discussion turned to attachment theory. A wealthy tech entrepreneur stood up and told me that he currently had four rescue dogs — all of them German shepherds with troubled pasts and behaviors ranging from aggression to anxiety.

What, he asked, did attachment theory offer to explain his love of these dogs and his need to rescue them? His wife was always asking him why he had these dogs. What could he tell her?

I said I was not a psychologist but I had encountered this many times before. I would guess that his mother was cold and aloof, and that he had been an anxious, perhaps lonely child. His father, I would speculate, was remote and absent. When he encountered these beautiful but vulnerable and needy and endangered animals, he was replaying a scene, a living video, of his own sense of being abandoned and unknown.

He gasped, looked at me for the longest time, and then looked down at his wife. "How could you possibly know that?" he asked me. I hear it all the time, I said. And I do. I have had that same conversation with dog and cat owners hundreds of times, and I can almost unfailingly see a glimpse of the early emotional development that shaped the template of their life with animals, just as my life in Providence shaped my need for Lucky and my feeling for him.

Unless we understand ourselves, we can never really understand the animals we live with. They are so often a reflection of us. Every animal we seek, own, and live with is a reflection of a part of us, and can speak to us, if we only learn to listen.

It was not simply because puppies are cute

or because I am a magnanimous animal lover that I wanted Lucky so badly. It was really the other way around. I am an animal lover in part because of the human being I am, the joys and sorrows I have experienced, the things I need in life. Loving an animal is a selfish act, something we need, no matter how we like to sugarcoat our motives.

A trainer once told me that to have a better dog, I needed to be a better human. It was the best advice I have ever received about animals, and it is one of the foundational ideas of my approach to communicating with them. Accepting that idea is essential. It is the first step toward truly understanding animals.

Twenty years ago, a friend, an analyst, told me of a wonderful book by the famed British analyst Dorothy Burlingham, and while reading it, I suddenly began trembling, my eyes filling with tears.

Burlingham wrote in her classic book *Twins* about the child who feels alone and forsaken in the world. He creates a new family in imagination, builds a wonderful new life in his mind. Most often, this life centers on animals. "The child takes an imaginary animal as his intimate and beloved companion; subsequently, he is never separated from his animal friend, and in this way he overcomes loneliness."

This animal, Burlingham wrote, offers the child what he is searching for: "faithful love and unswerving devotion. . . . These animal fantasies are thus an attempt to substitute for the discarded and unloving family an uncritical but understanding and always loving creature."

It was Lucky who came to mind. He had never really left.

I was already up when the alarm started ringing at 4 a.m. It was bitterly cold. No one was awake in the house. I got dressed, tiptoed downstairs, made myself a glass of warm milk, and buttered a piece of white bread.

The thermometer outside the kitchen window read 3 degrees, and I could hear the bone-chilling wind rattle the windows of the house.

I was too excited to eat much. I couldn't stop thinking of Lucky, of having a pal, a companion. The idea of him just opened me up in a way that nothing else had.

It was a long, cold walk. I ran much of the way. I wanted to make sure I was first. I remember my frozen nose, fingers, and toes. When I got to the boys' entrance the bells from the big church down the street chimed five times. I knew lots of kids might want the puppy, but I didn't think many needed him more than I did. I had never known the sensation of wanting something so much. It seemed my heart would burst.

I took my spot on the stone steps and spent the next couple of hours dancing and running in circles to keep warm. I counted to twenty and back a thousand times, imagined that I was one of the Hardy Boys out on a dangerous mission, plotted my life with Lucky.

Lucky would sleep in my bed, of course. We would take walks down the street, into the parks, through the big cemetery on North Main Street; we would play in the back yard, hole up in my room together while I read. In the summer, he would come with me to the beach, to Cape Cod. We would swim together, walk on the dunes.

Lucky would love me, of course. He would offer me faithful love and unswerving devotion. There would be no need for talking. Without words, we would understand one another completely. I *had* to get this dog.

Around 6:30 a.m., a half hour before the school bell rang, a big sixth grader named Jimmy walked up behind me, took a look around, grabbed me by the throat, and threw me off the top step and onto the asphalt. I got up and yelled "hey," and he punched me in the nose and knocked me down again, blood spurting out of my nose.

This, I understood, was life on the playground. The people in charge never saw this stuff; usually they were too busy yakking with one another to care or pay attention. Anybody who ratted on anybody else would not lead a life worth living. Even more than not wanting to get beat up on, I didn't ever want to be a rat. It was the strange code of the embattled boy.

We were on our own, the two of us, and to make things worse, I started to cry. I saw my life with Lucky washing away in blood and tears. I felt hopeless and desolate. Jimmy no longer bothered to even look at me; he just snorted and said, "That dog is mine."

To make things even worse, about a dozen other kids had appeared behind me to get in line for Lucky. I was no longer even in line. I had a growing audience to my humiliation; it would soon be the story of the day in school and at recess.

But the fates intervened. There are angels here on earth and sometimes they do appear.

"Wait a minute," said a thickly accented

voice through a window alongside the boy's entrance door.

I was still on the ground when the door opened and Mr. Wisnewski, the janitor, popped his gray head out of the door. We rarely saw Mr. Wisnewski and never spoke with him. He was a mysterious figure who would bleed the radiators when it got too cold in the classrooms or replace the light-bulbs when they went out. He was the one who cleaned up when somebody got sick or the toilets overflowed. He also came into the big auditorium to sweep up after assemblies.

There were all kinds of rumors about Mr. Wisnewski, most of them suggesting that he was a World War II refugee of some kind. He was believed to live in a tenement in North Providence. It was said he had no family, but nobody really knew since a conversation with him was unimaginable. It wasn't that he couldn't speak English; it was that he never did speak it, and we all assumed he didn't want to. Children often avoid what is strange to them.

He looked angry standing in the doorway.

"You there," he said, wagging his finger at Jimmy. "I saw what you did to that boy. You took his place in line, you hit him in the face."

Mr. Wisnewski, in a fur-lined cap, wearing his green uniform, came down the steps and helped me get up. He took an old rag out of his pocket and gave it to me to wipe my tears and the blood off my face and shirt.

"You are first in line," he said. "You get the puppy." Jimmy skulked away, glowering at both of us. I had not heard the last from him.

Mr. Wisnewski took me in through the door and led me down to the basement. We went through the door marked janitor. On the desk was a cardboard box with a small white puppy about the size of a big shoe. He was thin, trembling either from being nervous or cold.

"Lucky," I said. I had given him a name the first time I thought about him. We were going to be lucky together.

Lucky, as I remember him, was small and bright white. He had big and soulful brown eyes, like a baby seal. He was, at that point, the size of a small teddy bear. He was easy to pick up, soft and warm to hold. He looks in my memory like a cute Lab puppy, although I am certain he was a mutt. I remember him from the moment I saw him as being everything Dorothy Burlingham suggested — my trustworthy friend, my unconditional love, my buddy and safe place.

Lucky stayed in that basement room all day. I ran to see him a dozen times. I brought him water and a piece of my sandwich, which he ate greedily. Mr. Wisnewski had thought to bring some kibble, which he doled out of a tin. It didn't occur to me for many years that the puppy may have been Mr. Wisnewski's. That might explain why he was so involved in the adoption.

Nobody mentioned the blood on my shirt that day, but the girls went wild over Lucky — more girls spoke to me that day than had spoken to me all year. There was a line all day to visit him and pet him. Even the usually severe Miss McCarthy came in to take a look.

I stopped on the way out to thank Mr. Wisnewski again for saving my place in line, for helping me get Lucky. He just nodded and patted me on the head. "Good luck with the dog," he said. "Watch out for those boys."

Then he picked up Lucky, patted him a few times. "Take care of him," he said. "You are both lucky boys."

I had the idea that Mr. Wisnewski knew what it was like to get pushed out of place in line and punched. It's a curious thing, but I don't believe I ever saw Mr. Wisnewski again after that day. I have no memory of him beyond that encounter.

It took me a long time to carry Lucky home through the cold. I wrapped him in my jacket, he was shivering so badly, and soon, I was shivering even more. I stopped in the foyer of the branch library — I didn't dare bring him inside — so we could both thaw out. Lucky grew heavier and heavier as my fingers grew numb, but I was nothing but happy.

When I got home, my fingers and cheeks were nearly frozen. I took Lucky upstairs into my room. I laid him on the bed, wrapped him in some of my undershirts. He curled up in a ball by my pillow.

I cried again, this time in relief. My magical friend had come. I was not alone anymore. I don't recall ever being so happy or lighthearted; my troubles and fears seemed to melt away. Lucky needed me more than I needed him, I thought, or perhaps as much. The world seemed to change for me in that moment; it had become warm and safe. Lucky, I knew, loved me without condition or judgment.

I sat with him for hours, until my mother got home from work, came up to my room, and dragged me downstairs to have dinner.

I carried Lucky downstairs in a box; we sealed off the kitchen and he was permitted to walk around in there while we ate. He whined for me most of the time. See, I told my mother, see how much he loves me? My mother said Lucky was cute, but she quickly reminded me of my promise to take care of him. I could tell she liked him. She said he looked thin, that we needed to fatten him up a bit.

My father came in to check Lucky out. He picked him up and let him lick his face. I was surprised, since my father did not care for dogs much; my mother was the dog lover. After the first encounter, though, my father had nothing much to do with Lucky. I don't recall him ever touching him again and he wanted no part of his care or training. I remember what a different world it was for dogs then.

But even then, Lucky was at the center of my emotional life. From the first moment, I talked with Lucky. In the school basement. On the way home. In my room. On the kitchen floor. I imagined him always to be talking back to me, responding, agreeing, supporting.

I confessed to Lucky that I was a bed wetter, that I could not sleep over at the houses of other kids because I was too afraid of having an accident.

I told Lucky every detail of the most painful experience of my life up to that point, the story of how I had to leave summer camp because I wet my bunk every night and the other kids were making fun of me. I was not allowed to say good-bye; my clothes and sleeping bag were collected and a counselor drove me in silence back to my house, three hours away.

I told Lucky this so he would understand if there was an accident in the night. Puppies, I knew, also had accidents. I imagined Lucky to be as happy to be with me as I was with him. My father, I told the puppy, was angry with me every time I wet the bed; he would come in to give me lectures in the middle of the night and I would pretend to be asleep. If he comes in, I said to Lucky, pretend to be asleep, too. He'll go away.

I had never talked about my bed wetting before, not with any living thing. It was a great relief to share it. Lucky, I noticed, didn't care. That night, for the first time in more than a year, I did not wet my bed. I slept through the night, Lucky there next to me.

Every night, Lucky curled up with his head right under my neck and we both slept together. Lucky did not have any accidents in the bed, either. Not one.

He was a calm dog, perhaps too calm for a puppy. He slept for most of the day, and while he loved to walk around the backyard, and inspect my room, he never zoomed around like other puppies did. I brought him old baseballs to chew on — he loved them — and I scoured the refrigerator for chicken and tuna fish leftovers to feed him.

He seemed very happy just to be with me; he was perhaps the first creature who loved me in that way, who showed me what that might mean.

We did seem to understand one another. I believed it then, I believe it now. We communicated in some mystical and intuitive way, beyond words, with emotions and feelings. I knew when he was hungry, when he wanted to play. I believed that he knew when I was frightened or sad. He always seemed to do something to cheer me up at those times, or so I convinced myself.

I brought him scraps from dinner — bread, some pot roast, a piece of apple pie. He loved it. He loved breakfast, too, especially toast and eggs. I got up early every morning to take him outside and feed him. I brought him old gloves to chew on, walked to the butcher near my grandmother's house to get him a big heavy bone (his tiny teeth barely scratched it). I rubbed his belly, sang him to sleep, hauled him outside to pee and dump twenty times a day. Even then, I understood that the best and quickest way to housebreak a dog was to give him few or no opportunities to go in the wrong place.

I was as good as my word: I took care of Lucky, cleaned up after him, swatted him with a newspaper like my father told me to when he had a rare accident inside the house, usually near the back door. The newspaper swat was a common method of correcting a dog's behavior at one time. It is no longer an accepted method, at least not by trainers. Now I know how to use a crate to housebreak a dog — simple and not in any way traumatic.

Intimidating a dog is not the same thing as training a dog. Using physical abuse to compel a dog to do something can lead to fear, anger, and confusion.

Lucky got me through a few of those long school days, through Jimmy's taunts in the recess yard, which didn't matter to me anymore. I had the dog now. I wrote essays about Lucky in English class, drew sketches in art class, went to the library and researched the history of dogs. When Lucky arrived, I started reading about dogs and I have never stopped. Later on, I started writing about them, and I have never stopped doing that, either.

Lucky had only been with me for several weeks when he began to grow frail, and I saw right away that he was sick. He threw up his food, started having accidents. He had recurring diarrhea, his eyes were rheumy, and he seemed to weaken. At night, I would sleep downstairs in the kitchen with him. My father brought a mattress down for me to lie on. Lucky wasn't moving much by then; his breathing was slow, labored. I had a dream one night during his illness. *Let me go,* he said. *I am sorry, but I have to leave.*

When I got home from school one day, I couldn't find Lucky. There was a note from my mother saying he was sick and he had gone to the dog hospital.

I went up to my room and cried all night. I knew the minute I saw that note that I would never see Lucky again. It was as if he had told me himself. My miracle was over; the nature of my life had reasserted itself. My heart dropped right through the floor. The walk to school the next morning seemed like the longest walk I had ever taken. I must have looked awful. Even Miss McCarthy stopped me in the hallway and asked me if I was okay, if I needed to go see the nurse and lie down.

In the world I grew up in, children, like dogs, lived on the margins. I was presumed too fragile and innocent to understand what might befall a dog. Ignorance was protection in some twisted way.

My parents would not take me to see Lucky, and they would not tell me where he was. They would not tell me what his sickness was or what, if anything, could be done about it. Every time I asked them about it, which was continuously, they said he was being cared for, that there was no news. I stopped asking.

The next Saturday, my father took me to Rigney's ice cream parlor on Hope Street and bought me a sugar cone with two scoops of black raspberry. He got a cone also, and we sat at our table, licking away at our cones. They tasted wonderful, even in winter. I had heard my parents talking softly downstairs on and off for days. I knew they were hiding the truth from me. Kids always know.

I waited.

"Listen, son," said my father. "I wanted to tell you that Lucky is very sick. He has something called distemper. It is very serious. Lucky has gone to a farm in Massachusetts where they will take care of him and maybe he will recover. He'll be happy there. He's not coming back."

My father said that was all he knew, all there was to say about it. The farm did not allow visitors. He was sorry. He knew that I loved the dog. Life is like that, he said. You have to get used to it, you have to learn how it is sooner or later and deal with it; this is a good time to start. Life can be rough; it can throw you curveballs. I asked a lot of questions for a while, but I never got answers or any new information.

Eventually, I stopped asking and Lucky was never mentioned again.

So much has happened between then and now, but bonding with and caring for Lucky, my intense early need for this small dog, was a transformative experience for me.

Lucky taught me my first lesson about the power of animals in our lives, what they can mean for us, what they can do for us. He also taught me never to lie to my daughter about what happens to them.

As powerful as it was, my experience with Lucky was primitive. It was my first step into the world of animals, only a shadowy view of what was to come. I don't know if Lucky and I were really communicating. It felt that way, but how can I know looking back after so much time?

I cannot tell you one thing my teachers taught me in that school, but I can tell you every single thing I felt and learned in my time with Lucky. It is as sharp and fresh as if it had happened yesterday.

When Lucky left, I started wetting my bed again, but my life was not the same as it had been before. I was not the same. Once you have experienced love, trust, and companionship, you know it's out there, even if you can't always find it. Even then, I believed he had come to me with a purpose, and for a reason.

Lucky showed me there was feeling beyond my isolation and loneliness, love beyond my sadness and fear. He came to show me that I had more strength and resilience in me than I realized. After all, it was me who got the dog in the end. I always believed that was really his purpose in coming to me, to show me what was possible, even in a schoolyard, alone in the dark, surrounded by bigger kids — that in the face of all that, I could do it. I could get the dog. And I did.

2
VISUALIZATION
PICTURING THE FUTURE

After Lucky, I drifted away from dogs for the rest of my adolescent years. I wouldn't have another dog again until I got married in my early twenties, when my first wife moved to Washington, D.C., where I worked as a reporter for the *Washington Post.* This was a time in my life when I was focused on ambition and my professional life. I moved from city to city — Washington, Boston, New York, Dallas, Baltimore — eventually transitioning from newspapers to television.

Still, the dog we had with us during that time wasn't really mine — she ultimately belonged to my wife. Other than the occasional walk, she wasn't a factor in my chaotic, mobile life. Bean was a rescue mutt from a shelter in Virginia. Smart, loyal, energetic, she followed my wife everywhere. I was very fond of her, but we did not connect in the powerful ways that were to become a trademark of my life with dogs and animals. I suppose it's like romantic love in that way; you find it when you are ready.

The first transformative dog of my adult years was Julius, who came to me when I decided I could not survive in the corporate world and set out to become a full-time book writer. From Emily Dickinson to Virginia Woolf to E. B. White, writers have loved and needed dogs. They have served as creative muses, sounding boards, and companions in a solitary craft.

In my mid-thirties, I had become something I always wanted to be. I wrote about media for magazines. I wrote mysteries, novels, and nonfiction books. No more offices, bean-counting bosses, traveling, racing to cover stories, managing newspapers, or appearing on TV news shows. I had come home to myself, and although I didn't know it at the time, animals were about to become the foundation of my work and my life.

Although I had spent years in offices and around people, my headquarters for the next decade or so was a dark basement in an old colonial in media-centric Montclair, New Jersey, home to writers, academics, *New York Times* editors, TV producers, hotshot agents, and journalists.

It was a big change, a big risk. I was out on my own. I knew I was going to be alone every day, beset by distractions and temptations. I needed to learn the art of discipline, of motivating myself to get up every morning, get to my new, flashy Apple computer, and write every single day. I decided to start every workday with a walk around my pretty tree-lined block. Montclair is a beautiful suburban town, filled with gracious old homes and wide streets. I thought walking might clear my head, get me set for the working day.

I never really felt comfortable in the Jersey suburbs, though. I remember feeling odd everywhere I went. I remember being lonely. Although my marriage was beginning to fail, I had no sense of it, no awareness of what was happening to me and around me. But even in the midst of all that uncertainty, I also remember the wonder when I felt a warm nose at my side and looked for the first time into the big, brown, gorgeous eyes of a yellow lab named Jade.

I had been walking alone when Jade came running up to me as if we'd been friends for years. She was full of joy, her tail wagging, eyes wide with love and greeting. Her owner came running up apologetically, but not in alarm. She knew Jade wasn't going to harm or frighten me, or run off.

Caroline was a well-known artist and poet; I knew her name right away. She told me her dog Jade ran with her every morning. I was surprised to see no leash or restraint.

"She stays with you like that?" I asked, incredulously.

Caroline told me that Jade was an integral part of her work and daily routine. She kept her company, staved off loneliness, and helped ground her work by running, walking, and sitting with her. Jade was her inspiration and her companion. She could not imagine her life without her.

This was something of a shock to me. I had never had a dog like that or really heard anyone speak about that kind of relationship with a dog beyond Lucky, but I was very young, and she had not lived long. I did not know there were dogs that would stay with you in a congested town with no leash, or sit with you as you worked, or walk with you as you tried to organize your mind to think and write.

The idea that a dog might sleep in bed, be cremated at death, or go on vacation was simply beyond imagination or reason. Our dog Bean, who had died a few years earlier, slept downstairs on her dog bed, never came on vacation, and had nothing whatsoever to do with our work.

Jade awakened me to a completely different vision of animals. I was mesmerized by the idea that this sweet, beautiful creature — something about Jade was soothing and so grounding — was a creative partner, a signpost, a companion. Caroline gave me the name of Jade's breeder, who lived in Ramsey, New Jersey, and I went out to meet him. Six months later, I went back to Ramsey and came home with the most adorable puppy I had ever seen. He was small, mellow, snow white, soft, and sweet. Even then, he was easygoing.

I named him Julius, my father's middle name. He cost six hundred dollars; my grandmother would have fainted dead away. In the car on the way home, Julius put his head on my lap and went to sleep. I could feel us bonding even in that first hour. We were fusing, Julius and I. I had entered a different space, a different universe. I had my first lifetime dog, my heart dog. Julius loved everyone in the world, but he was, from the very first, my dog.

On my morning walks, he came along, and after three months I took the leash off him; he never once in the years I owned him strayed away, chased another dog, or went into the street without permission.

Julius enchanted the entire neighborhood; he was adopted by every child and nanny and mom and dad. I had to walk him early or late, because our walks took a long time, since he had so many admirers to meet and greet. His tail was spinning like a rotor blade from the minute we left the house to the time we returned. We often greeted the school bus that stopped in front of our house, and Julius was thronged by squealing kids lining up for a hug or pat. He got quite a few leftover bits of sandwich, too.

Our walks became a seminal part of my work. During them I would often speak aloud of new characters, story ideas, and plots twists. Walking is one thing; walking with a dog like Julius is something else, a totally different experience. I found it inspiring and comforting. He never strayed too far ahead of me or too far behind. He paid no attention to other dogs, was easy with people, and loved children to pet and hug him. I never had to worry about him, wonder where he was or what he might do. I trusted him completely. I walked with him in total peace and harmony, free to think about my work, to plan for the day. Whenever I looked over, he was right there, alongside me.

I worked alone, but I was not alone. I

worked in love and companionship. I had never experienced this before. It changed my work for good. Julius changed my life, just as Jade had done for Caroline.

There was an athletic park near our house where Julius and I made the rounds early in the morning. Every book I wrote in those years, every column idea, was hatched on those walks. After the walks, I'd make myself a cup of coffee and go to work. Julius would come downstairs, curl up at my feet under the glass table, sigh, and settle. As long as my computer was on, Julius never moved. He never barked, never stirred. He simply gave himself over to my need to work. When I got tired or stuck or frustrated, I would stand up and say "let's take a walk," and Julius would be up and out the door with me. We would work out the problem together. It was as if he understood what I needed him to do, to be. As if I had communicated the life I wished to have with him, and he accepted and embraced it.

Julius came to define my transition to life as a full-time writer. There was a special rapport between us, a communication happening through feelings, images, yearnings, signals — the first evidence of what I was later to call visualization.

Visualization is a daunting term, unnerving and confusing. I encountered it when a therapist advised me to use visualizations to get over the trauma of a divorce that ended a thirty-five-year marriage. I was hesitant at first, but it turns out that in the animal world, visualization is a simple process used frequently both by humans and animals.

Temple Grandin visualizes livestock traveling through pens. Linda Tellington-Jones visualizes a healing process for horses. Veterinarians, trainers, handlers, farmers, and wildlife experts use visualization to communicate with animals in the wild, from chimpanzees to elephants.

At its simplest, visualization is imagining what you want to happen. When Cesar Millan urges dog owners to "be the leader," that is precisely what he is telling them. Think about the future you want for you and your dog, and picture it clearly in your mind. If we form clear and consistent images of our desired outcome in our heads, animals will absorb that from our own body language, attitude, and even smell.

My first attempts at visualization were instinctive and primitive, and I was wary of even trying. I began, as always, with food. During those first few months, I put liver treats in my pocket; sometimes I gave Julius one, sometimes not. But he always paid attention to me, watching me as I began to walk. If he was calm around a dog, I threw him a treat. If he veered toward me and not the road, I tossed him another. If he lay quietly at my feet while I wrote, I tossed him another.

And here's the most interesting thing: if I wrote something penetrating or beautiful or thoughtful, I threw him several more. After a few months, he was sending me images and inspiration. I received them as impulses, beams of light, flashes of feeling.

It didn't really matter to Julius after a while whether I had treats. He associated them with me, accepted verbal praise happily, and it became completely instinctive and natural for him to stay close to me, to be calm. Sharing my workspace became his instinct. I noticed that the Apple "gong," the soft bell that sounds when the computer starts up, was Julius's cue to gather himself and settle down by my feet.

On cold winter days I sometimes used Julius as a footstool, resting my feet on his big, soft belly. On dark days when writing was hardest for me, he brightened the room and the space, his presence and the sound of his breathing and sighing gave it a warm, even cozy, feeling. On warm and muggy days, I turned on the AC — Julius loved air-conditioning — and my writing space became a cave, a place of refuge insulated from the world beyond.

Every morning, I imagined how our days together would go before they unfolded. I thought of our walks, our time together writing, his love of children, his calm around dogs, his wondrous temperament and generous spirit.

Before we walked, I closed my eyes and cleared my head, moving into an almost meditative state. I saw an image of what I needed, what I wanted, and lo and behold, so it came to be.

I got the dog I needed and wanted, the one I never imagined or knew existed. It was wonderful to me to feel this connection and think about it. At the time, I knew little about dogs, I avoided the growing shelves of dog books in the bookstore, I thought the dog love I saw often seemed excessive, overly emotional.

My life with Julius was in my head and his. It was not about training or obedience, but something deeper. This process of communication is about intent, not obedience. What do we want? What do they want? How do we wish to live with our dogs? How can we show them?

Dogs are one of the few species that have

figured out how to live with humans, how to connect with them emotionally, to project loyalty and unconditional love. They have made themselves emotionally necessary to us, even if we no longer really need them in order to live.

They — unlike raccoons, for example, or squirrels — have done this by grasping what we need and offering it to us. This makes them especially receptive to our intent. They want to please us. They want to know what we want from them. It is their instinct, their means of survival.

As I've explained, food is not simply a treat or training tool for animals; it is the essence of life itself, their purpose and focus and instinct and drive. Animals live to eat. Dogs love us so much because we feed, care, shelter, and pay attention to them.

At first I would instinctively toss Julius a treat whenever he was behaving like the dog I wanted or needed. Then I would reward him occasionally — both with food and praise. Then I repeatedly visualized what I wanted him to do. I sensed from the first that he received visual messages even when he could not understand the verbal ones.

I would pause, gather myself, clear my head, and picture what I wished him to do. *Stand quietly. Lie down in my office. Eat calmly. Approach children slowly and gently. Walk alongside my left knee.*

Sometimes, unfamiliar images would appear in my head, and I came to see that they were coming from Julius. This, I had learned, was how dogs talked to one another; this is how we were beginning to talk.

Because Julius was a grounded and healthy dog bred to be with people and stay still for long periods, it was already his nature to stay close to me. In a sense, our walking and working together was a replica of the hunting experience Labs have been bred to undertake for hundreds of years.

I am not a hunter, but the meaning of his behavior did not escape me. Were we out in the field, Julius would have naturally sat by my side for hours while I watched for ducks and geese. Lying at my feet while I typed, Julius was essentially doing the same thing. It was a behavior I could build on. Every dog has tendencies and behaviors we can build on, if we look for them and listen.

We often seem to be at war with our animals; our culture sets it up that way. In our world, a good dog is a dog who abandons most of the natural behaviors of a dog — having sex a lot, digging holes in yards, running off after strange smells, eating revolting things, fighting with other dogs, chewing up garbage, pillows, and table legs, stealing food off counters. In the natural world, this is simply what dogs do. To us, this makes them "bad" dogs. We scold them, reprimand them, rub their noses in urine, swat them with newspapers, shout at them in words they don't understand.

Vets, trainers, behaviorists, and breeders will all tell you that dog training in America is a catastrophe, and an expensive one. Millions of people are spending a lot of money to learn about things they will never be able to do and that will most likely not work for them.

A friend of mine was having lots of trouble walking a small dog — a Boston terrier — through the streets of Chicago, and to a dog park close to his apartment. The dog had arrived with a new partner, and was not originally his. This dog was battling him every step of the way, freaking out in the elevator, running in circles on the sidewalk, pulling and tugging at the leash. Walks were a nightmare, and an embarrassing one. In the dog park, the little dog nipped at people and ran in terror from the other dogs.

My friend knew of my work, had met my dogs, and asked my advice. I told him there was a great failure of communication between him and the new dog in his life. He didn't love this dog, I said, at least not yet. He and the dog were on different paths, leading different lives. I suggested he start communicating with his dog, listening to him. He was annoyed and uncomfortable with the dog's behavior: that was the message the dog was getting, and the dog was talking back, saying loudly and clearly in his own language, "If you don't like me, I don't like you, I don't have to do what you want, I don't know what you want."

Most important, I said, you need to begin imagining how you want this dog to be with you — calm, quiet, responsive. My friend said he realized that he didn't love the dog yet, and was annoyed by him. He understood this was his problem, not the dog's. He'd always had big, easygoing dogs, never a dog quite like this. He admitted he had a stereotype of little dogs in his head; he thought they were yippy and annoying. Since that is what you expected, I said, that's what you got.

My friend calmed down and took things step by step. He stopped shushing the dog in the elevator, and instead let him get comfortable with the crowded space. He acknowledged his own impatience and stopped pulling and tugging at the leash; he stopped rushing the dog and urging him to pee and move along. He let the dog alone in the dog run to work out his social issues, find his place in the pecking order, figure out what he needed to make himself safe.

In just a few weeks, a dramatic change occurred. The simple realization that he was projecting all kinds of troubling images to his new dog altered the dynamic. Bit by bit, the dog calmed down. He was still yippy and more anxious than a Lab, but less disruptive each week.

My friend came to appreciate the little dog's intelligence and lively personality. As he relaxed, the people around him relaxed with the dog, and appreciated the dog more. The dog responded to this. He came to see his new environment as safe and interesting, and he began to forge a new relationship with my friend.

"It is very different," he told me after a few months. "Our walks are calm and easy. I'm liking him more. I will be loving him soon. I am a dog lover, after all."

Human behaviors block communication with dogs all the time. Many of us forget that they are animals, not children or people. We project our own emotions and neuroses onto them. We use words they do not understand. When we experience conflict with dogs, our heads are filled with unhelpful emotions — anxiety, frustration, anger. We do not have a clear image in our minds of how we wish them to be.

I know a man who is obsessed with getting his Lab to exercise. He has turned the dog into a ball addict. He says he wants the dog to be happy and fulfilled, but this approach is misguided. Chasing a ball is a hunting activity, I told him. In excess, it is neither fulfilling nor happy; it's simply arousing. It's not necessary for the dog; it's necessary for the human to feel good about the dog. And it is good for pet stores who want to make money on accessories. For most of their long history with humans, dogs did not have or need balls or toys; they made do with the world around them.

This distinction — joy versus arousal — is central in learning to understand our dogs and pets as animals. Dogs love pack activity and exercise, but very often, what people perceive as fun and play is actually a hunting behavior that they have inherited as the predatory descendants of wolves. A Lab chasing a ball is not playing; he or she is exhibiting prey drive. Too much ball chasing turns them into obsessive hunters, not happy playmates. There is a very fine line between arousal — what a predatory animal feels when its hunting instincts are kicked into overdrive — and relaxed playing. Many people don't understand the difference.

With my dogs, I've learned to restrict hunting activities. We "play" outside, for fifteen or twenty minutes at a stretch. We never play inside the house; that is a place for quiet.

Dogs in the wild sit around or sleep for about fifteen to nineteen hours a day, like lions. They don't need to be aroused all day with playgroups, balls, Frisbees, and human wrestling mates. Labs are wonderful dogs, bred to sit still in marshes all day, but many of the ones I see are obnoxious, aroused, and often out of control.

Understanding the emotions of animals is also essential to understanding whether and how they might grieve. Dogs do not understand the concept of death. When their canine companions leave, they have no way of knowing if they are down the block, on a walk, at the vet, or dead. Because they are pack animals, they will very much notice when a member of the pack is gone, and can even become disoriented or lethargic, but I have learned to be cautious about ascribing those symptoms to human-style grieving, which can be long-lasting and devastating.

Dogs are immensely adaptable. Thousands

of dogs displaced by Hurricane Katrina in 2005 were rehomed all over the country, and hundreds of thousands of rescue animals are rehomed every year. Very few of them suffered from prolonged grief and lament. We love our dogs so much that we want them to be like us, but we are not always reading their behaviors correctly. They have emotions, but they do not have *our* emotions.

I explained this to the man who thought physical exertion was the key to his Lab's happiness, and I gave him a mental exercise: Instead of taking the dog out to play, imagine him lying quietly in the house while you go online or read a book. Imagine him doing nothing. Talk to him when he is still, not racing round. Close your eyes and signal him with images of peacefulness and companionship. Imagine the dog you really wish to have and show these images to him; he will receive them over time.

Talking to our animals is possible, but not simple. Imagining what we want is a process in which the human consistently imagines and visualizes the behavior he seeks from his or her dog, and, over time, the dog senses it, understands it, and then internalizes it. This is what happened with Julius, who walked with me sans leash and never went into the street, approached children gently, and lay down at my feet quietly when I went to work. At first, these were learned behaviors, my hopes and intentions for the dog. Over time, they became his intentions and behaviors. That's the pot of gold at the end of the rainbow, when the process really works.

In the spring of 2014, Maria, my second wife (we were married in 2008), and I decided to lamb, something I had done a half-dozen times before, but something we had never done together. A farmer brought one of his rams, and five months later, our six ewes began to give birth. While a few of the births were tricky, we had the most trouble with Ma, a sweet, old rescue sheep who struggled painfully in her labor.

We love Ma. She's big, ungainly, and clumsy, but unfailingly friendly and curious. We imagined her lamb would be great fun.

I came out one morning and saw Ma lying on her side. Her water had broken but the lamb had not appeared in the birth canal. Ma's breath was becoming irregular and there was some awful bloody fluid coming out of her uterus and her nose. I decided she should be put down immediately, not suffer any further.

Maria was away. I called the local large animal vet and she came over quickly. She agreed with my assessment but suggested we try one more thing. She set up an IV drip for Ma with some antibiotics and glucose for energy.

The vet left, and Ma revived for a few hours, then began to struggle again. I started to call the vet back, then stopped, and stood still for a while. I cleared my head of anxiety and images of Ma dying. I sat down on the ground with her. I was quiet for a few minutes, and then I imagined her giving birth, saw her lamb's head emerge under her tail, saw her gathering her strength for a push that would bring her baby into the world.

I waited — I don't know for how long — but when I looked down, I saw the lamb's hooves beginning to emerge. I put on my surgical gloves, reached into the uterine canal, got a good grip, and pulled. From the other end, I felt Ma pushing and groaning. Suddenly, the lamb slid right out and onto the ground.

It was a startling and beautiful sight; it seemed miraculous to me. The lamb slithered out, covered in amniotic fluid, its eyes closed. The ewe frantically licked away the film from her baby's eyes and mouth, so it could see and breathe. For five minutes, the lamb struggled to balance and shivered, until it finally stood up. Red watched closely. When the lamb stood up, all of us took a breath — dog, mother, and human.

Poor Ma could barely breathe, but she struggled to her feet and began licking her lamb, cleaning him up, bonding with his smell.

I called the vet and told her what had happened. "There was no way that baby came out by himself," she said. "Whatever you told her, she heard and believed."

I think Ma did hear me. I sent her a powerful and very distinct image of a lamb moving through the birth canal and out into the world. Ma had answered me in the Pole Barn that day. But she didn't just listen, she told me things. She told me that she wished to live, she wished to have her baby, she did not wish to be euthanized. What I saw in my head, what I heard from Ma — all of it came to be, right in front of me.

It was one of the most wondrous experiences of my life to see that baby open his eyes, struggle to stand; to see Ma clean and comfort him, and bring him to life. I stood to take a breath, but then saw something moving in Ma's stomach. I reached in and pulled out a female lamb. She had twins in there.

I am able to talk to animals like Ma because I respect them. I don't simply see them as piteous and dependent beings. I see them as sentient, with a consciousness and will of their own. Ma is not my ward; she is one of my partners on the earth. We were in that barn together, sharing in one of the seminal experiences of life. And we got through it together.

I didn't have to wonder if Ma wanted to have her baby, or could. All I had to do was ask and listen.

Two years after Julius came into my life, I went back to the breeder in Ramsey and brought Stanley home. He was a Lab from the same place I got Julius. He was as intuitive, loving, accepting, and easy around my work as Julius. He walked just as closely and calmly, and loved everyone he saw just as much. From Julius and Stanley, I learned that the value of a good breeder is that I could select a dog inclined to live with me in the way I needed, rather than setting myself up to be in conflict with my dog, which was the case with so many other people I knew.

Julius and Stanley were the first to teach

me that there was, in fact, a way to a wiser and more mystical understanding of animals. We didn't talk in words and human narratives, but we talked all the time. I learned I had a gift for getting dogs to understand what I needed, and another for understanding what they were needing and feeling.

In 2000, these two dogs and I undertook our greatest adventure, and communication was a big part of it. I bought a cabin on the top of a hill in the Upstate New York town of Jackson. I spent a year on that mountaintop with Julius and Stanley. My time there gave me the strength and clarity to begin a great change in my life, to act upon a determination to live a life of meaning and spiritual depth.

At the time, I was approaching fifty, and my life was about to change in almost every conceivable way. I was unhappy with my life in New Jersey, and my first wife and I had begun to drift apart. Increasingly I was drawn to the country, pulled by the idea of learning more about animals and writing about them. I found them healing and challenging, and I had a deepening spiritual connection with them.

Up on the mountain, I wrote my first book about dogs, *A Dog Year,* and the focus of my life began to change, though I had no idea how much. Animals led the way for me. They were the magical helpers Joseph Campbell describes when he writes about the pilgrim setting out on the hero journey. You get lost, wander into a dark place, and sometimes, if you are lucky, you are found and guided.

In our year together in Jackson, I also wrote *Running to the Mountain,* a record of my movement toward a spiritual life with animals (Julius is on the cover of that book). My connection with Julius and Stanley deepened through weeks alone, walks through the deep woods, and evenings by a roaring fire. I saw that the connection I had with these two dogs was genuine, trustworthy, and transferable.

Julius and Stanley grasped my life on the mountaintop as completely and intuitively as they had grasped it in New Jersey. They figured out how to be calm when the coyotes howled, how to navigate blizzards and snow, how to walk miles in the deep woods without markers or trails.

These might have been difficult things for city dogs to adjust to, but they were not farm dogs. I had stumbled — accidentally, really — into a way of talking to Julius and Stanley. Looking back, I see that my ideas and techniques for communicating with dogs were quite primitive. Over the next two decades, they would be tested in ways continuous and far-reaching, mostly with dogs and cats, but also with other animals, especially those that have been somewhat domesticated and live near and around people.

I could not credit Julius at the time with the importance of his role in my life, but it has been profound. Many years later, I took Julius's ashes and scattered them behind the cabin where he and I spent that wonderful year together. Finding my voice as a writer, my rich and meaningful life with animals — none of it would have happened without him. None of it.

3
ORSON AND THE RINGS OF FIRE

Orson, a troubled border collie show dog from Texas who arrived in 1999, was one of the first animals who shaped my philosophy.

Orson was a shooting star. He sailed right into the heart of my muddled life and set it afire. Orson was the catalyst dog — the dog of love, memory, awakening, controversy, and pain. He was my best dog and my worst dog.

At the time he arrived, I was working on a book about technology and writing for *Wired* magazine and *Rolling Stone.* I was a media critic, and all sorts of new media were erupting. I knew the rise of the Internet was important, but I just wasn't that interested in it. It lacked the romance and grit I wanted to explore in my writing. I was drawn to stories, not trends. I found the media world cold, corporate, and confusing.

I was also restless in other ways — living

in a New Jersey suburb, up to my neck in Boomer kiddie culture. I was in my late forties at the time, living with my first wife, swimming in the Baby Boomer stream. I spent hours carpooling, driving my daughter, Emma, to guitar and piano lessons and playdates. The talk in town was about teachers and schools — which was the best, how to get in — and soccer and lacrosse. The grass in the parks was torn up for lacrosse turf, plastic and durable. The parks now smelled like airplane glue; the dogs didn't even want to pee there.

I had slipped into the very life, in the very place, I had always meant to avoid. The Boomer life, SUVs, perfect kids, the very best colleges looming always over the horizon. To our eternal shame, we started working on those admission letters early on, stacking our lives and the life of our child with lessons and credentials. You had to plan early, that was the mantra. I pined for escape, driving upstate whenever I could, starting to work on my books, holed up in my freezing attic to write. I wanted something else, but I was not sure what.

My marriage to a very good human being, my first wife, was failing. We had begun to fall out of love with one another, our lives moving in different directions. We were, as the cliché suggests, drifting apart.

My yellow Lab Julius had died. Stanley, my other Lab, was diagnosed with congestive heart failure. The vet said he would not be far behind his friend.

I began rooting around online, emailing breeders, making inquiries, and checking out dogs. It was as much a matter of passing the time as it was anything else. I wasn't really sure what I wanted or when.

One morning, I awoke to an email from a Texas breeder, Deanne. She had a border collie, she said, that might benefit from a new home. The dog had problems, she conceded up front; he had been raised as an obedience show dog, but that had not worked out. She had read one of my books and had the sense I might be good for Orson, that we might match up. Good breeders do that; they are essentially matchmakers.

Deanne was honest. She told me right away that Orson had issues.

She was not lying.

When I met him, it was instantly clear that Orson was in no way an obedience trial dog. Obedience trials are contests in which the most obedient dog wins a ribbon. The dog that sits on command, lies down when asked, stops and stays until released, waits until told to move, and obeys voice, hand, and even whistle commands has a shot at a blue ribbon. Orson, Deanne told me, had broke loose during one trial, tore around the ring, and drove off some of the other dogs. He disrupted several trials, wreaking havoc, and was booted out of almost every competition he participated in.

I wondered what Orson's owner might have been thinking. Deanne said that Orson's owner continued to care for him but had stopped taking him to trials. The dog was now confined to a kennel and did not have much of a life. Deanne wanted more for him. She had read *Running to the Mountain* and was touched by how much I loved Julius and Stanley. Might I be interested in Orson? Her question sparked several months of agonizing and uncertainty.

I knew enough about border collies to know that at that point in my life I shouldn't have had one. They needed to work, and I had no jobs to offer a border collie in suburban New Jersey. Nor did I know much about training dogs or working with them. This was still at the beginning of my life with dogs. It was the 1990s, and while I had had good luck with training Julius and Stanley, animals were not yet the focal point of my life. I was inexperienced with complex dogs and had never set foot on a farm or been near a donkey or sheep.

Border collies are almost all complex dogs, but this one sounded like he had more problems than most border collies, known for their intensity, restlessness, and arousal. Julius and Stanley were gentle, easygoing Labs, easy to train, and happy to lie around for the long periods of peace and quiet a writer needs. Border collies are not like that.

Still, the story of Orson nagged at me. He was stuck in a kennel in Texas with no life. A New Jersey suburb with a human who worked at home all day and had a fenced-in yard might be better. And then, there was this: I was becoming very drawn to the world of dogs — to training them, working with them, living with them. Something about them stirred something inside me. I wanted to know more, experience more, write more about them.

A challenging dog, Orson could perhaps teach me more about dogs, force me to learn more than I had needed to know to live with Julius and Stanley. Also, as a lost boy myself, I simply identified with his story. I knew what it felt like to be unsuccessful and alone and abandoned. I had spent a lot of years of my life feeling that way, all the way back to Lucky. Here, perhaps, was a chance to finally write a different ending to that story. It was as important to me in my forties as it was when I was eight.

Pining after a beautiful dog in trouble was an old story for me, and here was a beautiful sleek black and white border collie with radiant brown eyes and tons of personality (Deanne sent me photos of him online) looking for a home. I was a person in search of a purpose.

Deanne and I talked on the phone for hours, emailed one another a hundred times. I fussed back and forth in my head. Was this a good idea? A good way to get a dog? The right way to get a border collie?

Ultimately, I said yes. I can't honestly and fully say why. I have a long history of impulsive and unpredictable moves, many of which have worked out for me. But this was not the way to get a border collie. I knew that then; I know it now. But I did it anyway.

A few weeks later, after elaborate negotiations and preparations, Orson was shipped from Texas to Newark International Airport. I was excited about Orson's arrival. I read a dozen books about border collies — most of which terrified me — and got several crates, repaired patches in the backyard fence, and stocked up on dog food and throw toys and balls for exercise.

I got to the airport early, found a parking spot close to the terminal, and stood eagerly under the arrivals sign. I had an enormous sense of excitement and expectation.

I imagine that the old-timers who work at the airport are still talking about Orson's arrival at Terminal C. I am. Baggage handlers dragged his crate off a truck and hauled it into the busy terminal. I rushed over, leaned down in front, and slid open the front gate. His arrival was well before 9/11, and I am grateful for that, because things might have turned out very differently for us if he had come a decade later.

The next thing I saw after I opened his crate was the terminal ceiling. Orson shot out of the crate as if fired from a cannon, knocked me on my butt, and ran into the crowd. I knew he was beautiful, I saw that much, and I knew he was strong and fast. When I got to my knees I saw him vaulting over one moving carousel after another, to the shouts and screams of dozens of surprised travelers.

You have to consider the scene from the point of view of an excitable border collie, plucked from his home, driven to an airport, stuffed in a crate for a five-hour flight, tossed onto a moving cart, dragged into a huge building with thousands of people milling around — and suddenly his crate door opens.

He is out of there, like a missile fired from a silo, into the mayhem of a busy airport terminal at a peak travel time: there are flashing lights, honking carousels, moving luggage, enough things to make a border collie frightened, crazy, and excited. Plus, I think Orson was all of those things to begin with.

I was suddenly in the middle of a genuine nightmare, a combination Marx Brothers fiasco and horror movie. Orson rushed from one end of the terminal to the other, with me and several Port Authority police officers in hot pursuit.

Some people screamed at the sight of this wolflike creature diving over the baggage conveyors, darting past strollers and between people, looking frantically for some open space. I was shocked and frightened, praying he would not find any.

Some kids cried, some travelers dropped their luggage, a few dog people tried to grab at him as he flashed past. His speed and agility were astonishing. He was moving faster than anybody could keep up with. He had the Port Authority officers running in circles, huffing and puffing.

If Orson made it through the doors, it was likely that we would never see him again. Newark Airport parking lots are vast, filled with cars and trams, and surrounded by access ramps, superhighways, and shipping terminals.

There were several border collie owners in the terminal that day. The dog Samaritans, as I called them, came running. They instinctively knew what was happening and wanted to help. We all knew Orson must not find his way outside, something he seemed smart enough to do. We found each other and agreed through hand signals and shouts to head for the exits, since Orson would eventually make his way for the light and open air. All of us were shouting for people to get out of the way, holding our arms out to head him off, yelling to one another with reports of Orson sightings in one corner of the terminal or another. It was a good strategy, since four or five of us were waiting when he exploded toward the electronic terminal doors.

At this point, I got my first good look at Orson and saw that his tongue was hanging down to the floor. He was wild-eyed, exhausted, clearly in a frenzied panic.

There were enough of us at this point to

form a semicircle, and for the first time, he slowed. He seemed exhausted. I lay down on the ground, held out a liver treat as a peace offering — he had to be hungry — and asked everyone to be still. The police officers kept the crowds back and it got quiet. For the first time, Orson focused on me, looked me in the eye — I knew this was important.

He sniffed the treat and did not take it, but he did look at me long enough for me to get his attention. He was focusing on me. I spoke to him softly and calmly, throwing one treat, then another on the ground.

Orson took one of them, and then stared at me some more. This was good, a human for him to connect with. I did my very first visualization without intending it — I imagined him crawling over to me, getting close enough for me to put a collar on him. And that is what happened. A few minutes later, Orson was on a leash and we were both sitting on the ground beside my car in the parking lot.

We were in the midst of a huge and crowded lot. Buses rumbled right by us, enormous jet planes roared overhead, the blare of loudspeakers and announcements was deafening, an electric train hissed over our heads. Yet, in a sense, we were alone. I put a bowl of water out and Orson gulped it down thirstily. I put some kibble in another bowl and he wolfed some of it down. Food, the first step in bonding. I was suddenly more than just a large man with a leash. I was life itself. I could see the change in his eyes.

I kept speaking his name, petting him on the head and shoulders. I continued talking to him, getting him used to my voice. Orson was wild-eyed, disoriented, and exhausted, but his panic had eased and his fearsome curiosity had begun to emerge. He watched the planes in the sky, the trains, listened to the buses, the sound of car horns. He was like a kid who suddenly finds himself in the middle of a three-ring circus.

Orson rode home in the front passenger seat. I put the window down a bit so he could stick his nose out. He was not frantic anymore. He seemed to be relishing his role as passenger. He was fascinated by the traffic, the lights, and the buildings whizzing by. He was interested in his new world.

Orson was an explosive animal, the biological opposite of Julius and Stanley. My suburban town of Montclair was the last place an untrained border collie like him should have been brought — as was pointed out to me for years by outraged border collie owners and lovers. I guess I knew that from the first. But I was willing to learn and do what I had to do to make it work.

I knew I had to restructure my life, and right away. We needed quiet and space. I started getting up at 5 a.m. to walk him. I scoured the town for quiet parks, with geese for him to chase, some space for him to run. He did not seem to know how to walk quietly, or on a leash, or alongside a human being. He shot out after cars, diesel trucks, and other dogs. He wrapped the leash around my legs and ran in circles.

My wife wanted nothing to do with Orson. She had no interest in walking him or being near him. This was to be significant in the coming months, as caring for Orson took more and more of my time, and this separated my wife and me even further. In a curious way, this new dog highlighted the distance growing between us, the separate paths we had begun to walk.

Would I have seen it without him? Would she? I can't say. But I see now that instead of finding something that would pull us together, I had chosen something that was sure to push us apart, although I did not realize it at the time.

Orson quickly became a significant part of my life. I found some creative places to exercise him in the early morning hours, when he and I haunted the streets and parks of the town.

The police in their cruisers came to know us. They would wave to us as they drove by, me grappling with Orson's leash. Orson loved the backyard, but not for long. Within a couple of days, he had dug a tunnel under the fence, and a neighbor called to tell me he was in her backyard, treeing the squirrels there.

Soon after, Orson pried open the nailed slats in the fence and squeezed out. I found him in the schoolyard down the street, chasing after soccer balls.

Then he simply jumped over the fence, and he took off trying to herd a big yellow sheep that looked to me like a school bus. I was exhilarated by Orson, but also frustrated and angered by him.

Orson's troubles in the suburbs mounted quickly. He would scramble out of doors, windows, screens, pull out of leashes, slip out of collars. He chased after small dogs and herded them, driving them considerable distances as outraged and panicked owners raced behind. He tried to herd strollers over the objections of enraged nannies and hysterical moms and dads.

One afternoon, he dug under our fence,

traveled a mile or so, and dug under some-one else's fence, terrorizing the two toy poodles in the backyard, herding them into a corner by the side gate. The residents started to call the police when they heard the barking and shrieking, but then remem-bered they had heard about a crazy dog and his loon owner living several blocks away. Blessedly, they called me, not the police. "What have you done?" I asked balefully as Orson gazed out the car window on the way home.

I did not know what to do with him. Like so many dog lovers, I was drawn more to the rescue than the dog, at least at first.

One day Orson opened the front screen door and took off down the street. A neigh-bor came rushing over to tell me he was trying to herd a school bus picking up kids at the elementary school down the block.

When I rushed down the street to the school, I saw the principal holding Orson off with a stick as he tried to herd the bus filled with children; the driver had closed the doors. The parents screaming around the bus were not amused. I heard sirens in the distance. In my Boomer town, nobody had much of a sense of humor about dogs or children.

I grabbed Orson by the collar, picked him

up, ran through side streets, and hid in a friend's garage. Later, they smuggled me and Orson out and drove us to my house in their minivan. We stayed off the streets for days.

I did some unconventional things, like taking Orson down to the Garden State Parkway at 4 a.m. and letting him run after big trucks, which he loved. (He was separated from the trucks by a big long fence on our side.) Some mornings, I drove him to freezing and empty public beaches. I think I came to love Orson on a windswept beach in northern New Jersey.

One blustery and very cold winter afternoon, I took Orson to a national park where I could look out across to Staten Island and the Verrazano-Narrows Bridge. It was a beautiful spot, a beach, deserted and eerily empty.

Orson was a study in grace and enthusiasm as he tore along the shore, covered in salt and spray, barking joyously, confused by each wave's sudden disappearance, excited to see another one forming. For the most part, he stayed along the shoreline, but sometimes he got carried away and jumped into the water to try to herd a wave, to turn it. Stunned by the cold water, he'd retreat, then charge again.

He was determined and brave and powerful. It was a beautiful thing to see his spirit unleashed like that, and this was one of the very few activities that actually tired him, at least for a few minutes.

Afterward, I threw a blanket over him and toweled him dry, and he came and lay down next to me, tongue hanging down, eyes wide. We stared out at the big boats steaming in and out of New York Harbor.

We connected with one another, Orson and I. There was something in each of our souls that brushed against the other's. I was at peace there on that sandy beach; so, I think, was he.

I wish we could have stayed there, kept that feeling.

It was not so easy at home.

I was beginning to see that there was something wrong with Orson, something off. He had plenty of smarts and love in him, but there seemed no place he could really settle and be at peace. He was easily aroused, and his arousal often turned to borderline aggressive behavior — intense herding behaviors, a lot of nipping. (Recipients of the nips would likely call it biting.) He did not respond to training easily, even by professionals. There was a part of him that seemed beyond reach.

When we talk about dogs, we often speak of the "good dog" and the "bad dog." Why? It is difficult for me to think of a more distancing notion or an idea that is more counterproductive to real understanding or communication. There is no such thing as a "good" dog or "bad" dog, as the life of Orson clearly illustrates. Those are human constructs, marketing ideas that are meant to force dogs to fit into our ideas of life, ours and theirs.

When a dog enters the life of a human being, that animal is instantly challenged to stop doing many or all of the things that are natural to him or her: fighting, having sex, chewing up soft and tearable things, digging up a garden, pulling on a leash, barking at other dogs and passing trucks.

A "good" dog is a dog that gives up the identity of a dog and absorbs the identity of a human child, which is the identity most human dog owners seek. A "bad" dog is a dog that retains some of the strongest traits in the natural life of a dog — barking, racing around, squabbling, seeing many things as a bone, seeking to procreate.

The challenge of the human being is not to force the dog into unnatural behaviors that often leave her anxious and confused, but to learn to understand the animal and communicate with her well enough to help her understand what is required for her to live safely and contentedly in our world.

Orson reminded me that dogs are neither good or bad; they are, in the human sense, utterly neutral, without human ideas about right or wrong. When they are "bad," they are simply following their instincts and their nature. When they are "good" we have usually cowed or intimidated them into behaving in ways that we would like them to behave.

The "good dog/bad dog" myth causes us to live in conflict with animals who could and should be our partners, not our troublesome wards. It keeps us from talking to them or listening to them. It puts us in perpetual opposition to them, damages their true nature, and often reduces them to confused and unreachable children. To communicate with them, it is imperative to see them differently, in a new, more humane and knowing way.

You cannot communicate with any entity you do not understand, or who cannot understand you.

Dogs and other animals do not have consciences; they do not decide to be good or evil. These are concepts human beings project on them. We see them as mirrors, as reflections on us. Humans are the only animal in the world that thinks in terms of good and bad, and in our arrogance, we assume all of the animals we love must do the same.

The "good dog/bad dog" construct has done more to damage our ability to understand animals than anything else. This misplaced morality is as big an obstacle to communicating as our insistence on using words animals can't understand.

Within a few years, Orson had begun to deteriorate. He broke through screens, climbed through the gate, and pursued smaller dogs, frightening people walking with their children. He nipped at two or three people, and the sights and sounds of the suburbs, especially buses and sirens, drove him to a high and troubling state of arousal. I was not finding a way to calm him down. So I decided to find a sheepherding trainer and take him there to see if he could help.

I drove Orson up to the Tufts University veterinary school in Massachusetts for some tests, and a behaviorist there suggested he might have a neural disorder. I worked harder. I got more help — vets, behaviorists, trainers, neurologists, holistic practitioners. And I spent more and more money, thousands of dollars trying to unlock the secret that would help Orson settle down and live safely in our world.

I resumed our early morning treks through parks; he took to chasing geese with enthusiasm. I became creative looking for ways to wear him out, help him stay out of trouble.

It was hard. My neighborhood was densely populated, full of kids running, pushed in strollers, shouting and playing with one another. There were dogs everywhere, and Orson did not like any of them. He got too excited around other dogs; he barked and jumped and growled.

I saw that when he got too excited, he would nip and lunge, sometimes frightening people. It was a herding move, people told me, but I think I knew better. It was more than that. My growing panel of experts were in agreement: there was something wrong with him, most likely something neural.

Orson nipped at the leg of our mailmen, at the coat of a commuter heading home from the bus stop. He nipped at the ankles of a nanny walking with twins. She was frightened and said she would call the police if he ever came near the children she was caring for again.

His behavior was serious, troubling. He

was not an aggressive dog, he did not bite or attack, but his arousal and dominance were threatening to people, and it makes no difference to a small child if a big black and white dog is nipping or biting. It was both terrifying and unacceptable to others and to me.

And so my dilemma just grew and grew. The more I knew him, the more I loved him; the more I lived with him, the more trouble he caused, and the more uncontrollable he seemed. Was there any way out of this that would not be painful, or worse?

What did I learn from that dog?

I learned that I really knew very little about animals.

But I also knew that I could try to speak to Orson without words, even if I hadn't yet honed my techniques. As out of control as he was, as much trouble as he caused me, we had become very close. I could sense when he was about to explode, which could happen at any time, but he did begin to try to respond to my looks and feelings.

And that perhaps was what was so poignant about Orson. He did want to please, he did try, he did want to listen, but I grew increasingly certain — as did the experts I was consulting — that he was simply too aroused, that he was damaged in some way.

I began to visualize our walking calmly together, and that did begin to happen. I visualized periods of stillness and calm — a new idea for him — and that began to happen also. But it did not happen for long; any progress we made didn't stick.

I saw clearly that he did not respond to words or language, but then, he often didn't respond to anything else, either. And I couldn't experiment in a suburban environment the way I could later at my farm. The town was crowded, congested with dogs, cars, buses, and trains. Orson was living in the middle of an ecosystem that constantly aroused and unnerved him. Nothing I did worked for long, or seemed to change him for good.

One thing about visualization: as powerful as it can be, it isn't like faxing or emailing a photo. The person doing it has to be in the right headspace, and so does the dog. It isn't a magic pill, but a valuable tool. By itself, it wasn't enough for this dog. If I wanted to keep him — and I did — I would have to learn to enter his world. I soon realized that I would have to integrate him with sheep if I was to unlock the key to his craziness. I saw that working dogs need to work, not just run in circles.

If there was a single message he was sending me, it was perhaps this: *I need to find my center, to do the work I was meant to do, bred to do. I need your help in finding it.* I kept hearing. *Help me, please, help me.*

So I answered this call and took him to a sheep farm in Pennsylvania where a well-known trainer named Carolyn Wilki specialized in flipping troubled dogs.

Carolyn watched Orson, put him near sheep on a lead, gave him some commands. Her first recommendation was to change his name. Orson had come to me with the name Devon. Caroline pointed out that the dog got too excited when he heard "Devon," that he might be calmer learning to answer to a different name. If there were unpleasant memories from Texas, she suggested, a new name might also help with that.

So "Devon" became Orson, after Orson Welles. Orson he remains in my heart, and when I think and write about him, he is always Orson. I tossed treats at the dog and kept saying "Orson." It took about two days for him to consistently respond to the new name. But he did not calm down. Neither did I.

I learned a lot at the sheep farm, about training dogs, about living with Orson. Carolyn taught me calming training techniques, basic repetitive exercises that quieted the dog and helped him to listen and focus. Trainers often refer to these exercises as "obedience training," but I call them "calming exercises." They do ask the dog to obey some simple commands, but the point for me is not obedience. The point is to get the dog to a place of calm focus. It is a dialogue rather than a military-style training exercise. I don't want my dogs to "obey" me; I want them to understand what I am asking them to do. There is a big difference.

I went out to the farm two or three times a week. Orson and I did our calming exercises fifteen to twenty times each day, for five minutes at a time. We worked on lying down and staying, on eye contact, on him walking alongside of me. We ran around sheep pens, although Orson got too excited to herd the sheep — that would take a long while.

At the sheep farm, I was stunned to discover what I had been looking for all of my life. That deep craving that had no name now manifested itself as a life in the natural world, a life with animals, a life with dogs, a life free of the growing corporate strictures and suffocations spreading across the American workplace. Free of political correctness and the Boomer obsession with creating the perfect child. It was a life I was on fire to explore.

I went to Carolyn's farm, Raspberry Ridge, every chance I could get. I often slept over. Orson and I would get up before dawn and go to the sheep pen and open the gates. Orson would tear off behind the sheep and they would come roaring out and down the path to the big pasture nearby. He was great at getting the sheep moving, but he almost always got too excited to be left alone with them.

A grounded border collie will settle while the sheep graze, keep an eye on them without disturbing them pointlessly. Orson could not be still for that long; he simply became too excited. He ran too much, barked too much, nipped too much.

With some trepidation, I decided to enter Orson in a herding trial for beginners. Dog and trainer would come into the sheep pen; the dog was told to lie down, then walk up, then lie down. At that first trial, the sheep took one look at Orson and two of them jumped over the fence and ran off into the woods. Dave, Carolyn's resident working dog, took off after them and brought them back.

I was determined that Orson should make progress. We practiced it over and over again, week after week. *Walk up. Lie down.*

Stay. Sit. Simple enough, but not simple for him. Around the sheep, he became wild-eyed, drooled, and shivered.

Away from the farm, I doubled down on our calming training, chipping away at Orson's resistance, his distraction. I changed my voice, worked on body language, emotion. I imagined him lying down near the sheep a hundred times a day. I sensed he was eager to please me, eager for success.

We kept at it, day and night, in cold and hot weather; it was a full-time job. I remember sweating out there, freezing, getting stung by bees, knocked down by nervous sheep, gnawed on by flies.

There was at least one transcendent moment that occurred just moments before a herding trial. We were sitting outside the pen; a dozen sheep were inside. I closed my eyes and experienced an implosion of feeling, of light. I saw an image of Orson walking up to the sheep, lying down, and waiting for my command to move before getting up and leaving the ring. It was as clear as a video. It was coming from Orson, from outside my head and my consciousness. I knew he was talking to me, telling me he understood what I wanted, showing me that he could do it.

And so he did, exactly as the images suggested, exactly as I imagined. It was a fusion of two minds, a connection, a coming together, the result of feeling, hard work, persistence, and love. We had opened a channel to one another in that pasture on that hot June day, with dozens of people looking on.

Orson walked into the ring, trembling. The judge was wary. Orson walked up to the sheep, and when I said "lie down," he did, and the sheep moved to the other side of the ring. I told him to "come by," the herding command to run in a clockwise circle, and to "walk up slowly," as we had practiced countless times, and that is precisely what he did.

The sheep, sensing calm, did not bolt as they usually did. They walked around to the gate and then I told Orson to "lie down" and "stay" and he did. The sheep moved calmly around to the other side of the pen. There was applause from Carolyn and the other friends we had made.

We had done it. We won a blue ribbon. I was overjoyed. A new chapter in my life had been ratified, and perhaps there was hope for Orson. He had not had the makings of an award-winning dog, but that's what he was now. Getting him to that point required tremendous patience, communication, training. I learned that I was good at this, that I had a feel for it. Carolyn was impressed. She told me she hadn't been sure it could be done.

Surely, things would change.

I believe that Orson came to bring me a powerful message. It was this: you are also lost, you also need to find yourself, we will take a journey together and find out who we are and what we were meant to be. Orson changed my life, dramatically and irrevocably.

At Carolyn's kitchen table, I found other members of my tribe, lost and drifting souls looking for a purpose. I loved seeing their emotional responses to their dogs, studying the ways in which people and dogs attached to one another. I loved hearing their stories and watching Carolyn figure out how to train their dogs.

At the farm, I fell in love with the rituals of a life with animals — the herding, watering, the sounds and smells, the rhythms of life, the joy of working with a dog.

In the mornings, as the sun rose over the pasture and the flock settled in to graze, I would call Orson and he would sit down beside me, watching the sheep. It was a transformative experience for me, beautiful, stirring, and very powerful. I felt connected to the rich history and traditions of a life with animals. It was as natural as anything I have ever done in my life to be sitting out there in the pasture, the mist rising from the meadow, the sun glistening off the dew, me with my dog watching the sheep graze.

At moments, I could see Orson felt that, too. Despite all the chaotic craziness of his life and his brain, he seemed to grasp the meaning of those moments. He would settle, look at me curiously.

It never lasted long, and the chaos never really left him, but the experience of being with him had transformed me and my sense of animals.

Suddenly, in the midst of this intensive training, I had a new idea. I would sell my cabin upstate and look for a small farm — there were lots of them on the market at that time. I was fifty-eight years old, an age when many people begin to think about downsizing their lives. But I am not into downsizing; so instead, I upsized, big-time. I found a ninety-acre farm on the side of a hill in West Hebron, New York. I bought it, deciding to go there and write a book about my work with Orson and other dogs.

I felt that it was Orson who had led me to my new life, not the other way around. He was showing me that I was not happy in my life, not happy in my marriage. My wife and I were steadily moving in separate directions. She wanted no part of a crazy border collie, of a sheep farm, of life in the country. And I don't blame her; it was not her choice, not her life. It was selfish on my part, but I believe it was a matter of survival for me.

I was restless in my work, not writing from my heart, not following my bliss. I began to find it out there in Pennsylvania, on that sheep farm, just as Orson began to find himself. We did it together. Before the sun came up we came to collect the sheep, took them out to a nearby pasture, and guarded them while they grazed. We sat with them in the morning mist until the heads came up and their bellies were full.

What I was experiencing reinforced an idea that had surfaced with Julius and Stanley, two amiable and easygoing spirits.

Four years after Orson arrived, we were both living at Bedlam Farm, my new farm in Upstate New York. We were still struggling. A new behaviorist at the University of Pennsylvania told me he was certain Orson had suffered neural damage somewhere along the line. He was not trainable, the behaviorist said, and he was not curable. I was advised to keep Orson under close watch and confinement. I was told that my dog should be considered dangerous, that Orson could surely harm someone.

I tried the usual things with Orson, and some unusual ones — trainers, behaviorists, veterinary school specialists, holistic vets, Chinese calming herbs, acupuncture, massage, anti-obsessive medications. I tried calming training, sheepherding, tranquilizers, obedience work. I did this for three years, spending more than fifteen thousand dollars in the process. I was running out of money and realized I was losing perspective.

Orson had always had a problem with what one vet called "hyperarousal," but it worsened in the summer of 2005. He bit a painter working on the farmhouse in the leg. Then he jumped the fence and bit a UPS driver as he climbed into his truck. I increased the height of the fence and put Orson in the rear, in an enclosed space. He dug under the fence and nipped a carpenter replacing a window frame.

When I dug the fence deeper, he jumped over it. Inside the house, he had taken apart two metal crates. One morning, he chewed a hole through the mesh in the front screen door, raced into the yard, and bit a young gardener — a Bennington College student I had hired to help that summer — on the leg, drawing blood.

I gave Orson sedatives and confined him to two rooms in the house. When I was away, he nipped the foot of the dog sitter and ran out the door and into the woods. When I came home, the road was quiet. Orson was sleeping in the garden. I did not see the young boy bicycling down the road. He saw Orson and came over to the fence to say hello to him. Orson jumped up on a garden bush and lunged at the boy's neck, biting him severely below the chin. I watched in horror as blood spread over the boy's shirt and down to his belt.

A neighbor, a volunteer fireman, was driving by and stopped to help. He stanched the bleeding and may have saved the boy's life. We rushed the boy to an emergency medical clinic nearby. He needed ten stitches to his throat.

The boy's parents were friends and fellow dog lovers. They assured me that they would not sue, although I would not have blamed them. When it was all over, I sat with Orson in the yard. I was crying. He was shaking, sensing, I'm sure, the tension and anger in me.

Orson had been sent to me in the first place because of problems with arousal; his owner had found him untrainable. The behaviorist at Tufts sat down once more with me and told me Orson was damaged, that there was nothing I or anyone else could do. To confine a border collie like Orson in a small space for life would be cruel, he said. He would fight it every minute of his life.

I received a powerful message from Orson as we sat, devastated, in the yard. It came to me in the form of an image — Orson at peace, lying by a stream. A feeling, a sensation, accompanied that image. It was a message that I heard this way: *You need to help me leave. I do not know how to live in this world. You need to let me go. I do not seek to displease you, I do not wish to hurt any person. I do not know how to live in this world. I did my work. I led you here and brought you out of your unhappy life to a better place. I want to go.*

Receiving these messages isn't like hearing a spoken plea or voice mail; it's not like getting a letter or an email. The messages come from the love of an animal, from knowing him or her, from trading emotions, offering food, understanding instincts, from experience and acclimation and trust.

I had no doubt what it was Orson was telling me. And I think there was no doubt in Orson's mind about what I was telling him. I could never again tell myself or anyone else that I did not know what might happen, what Orson might do. I knew as I watched the blood run down that child's shirt that I would never forget that sight, and would never permit it to happen again.

Orson was not the kind of animal who could be easily or safely confined in a small space. Nor would that be humane. In the contemporary animal world, there is this entrenched idea of the "no-kill" policy for shelters. I can't see how it is noble or humane to confine a dog in a small crate for years and call it a loving thing to do.

My understanding of animals is very different. I can hardly imagine a crueler thing to do to a social and active animal like a dog than confine one in that way for years, for a lifetime. To me, this is the new and socially acceptable abuse of animals, this idea that they must never die, can never suffer, must live perfect and protected lives in the name of giving them rights.

Orson knew, I believe, that I could not abide what he had become. That it horrified and saddened me beyond measure, beyond words. Orson, like many animals who live with people, could sense human emotions, could smell them, read them. He understood what I was feeling. I knew I had to put him down, to take him out of this world. I knew it would be an unpopular and controversial decision. It didn't matter. The person I had to please was me; the ethics I had to honor were mine. I did not need or seek the approval of others.

Weeks after he bit the child, after long and expensive consultations with more vets, holistic and traditional, with behaviorists, trainers, and friends, I took Orson to the vet and put him down. Afterward, I carried him up to the top of the pasture and buried him.

I was attacked for putting Orson down by scores of dog owners and animal rights activists online, and am still occasionally called a monster and murderer online. There are whole websites devoted to hating me and my decision. But it was a good decision, and I am glad I made it. I would make it again. It was the right choice, the ethical choice. Before he bit that child, I could tell myself I didn't think it would ever happen. Next time, I would never be able to look a child or his mother or father in the eye and say that again. In my view, caring for dogs does not mean allowing them to harm people.

When Orson had been dead for several weeks, I was visited by a shaman who conducted a spirit talk with my troubled border collie. She described the place from which Orson had come, the place where he had returned: a green valley by a flowing stream, blue lights everywhere, Orson sitting in rings of fire.

His work was done, she said. He was ready to leave. There are lots of different ways to communicate with animals. There are people who can communicate with the spirits of dead animals. I am not one of them, but I know several, and I have no doubts about their powers. I have met animal communicators and animal shamans. I have talked with them, seen them do their work. There are many things in the world that I cannot fully understand or explain, but that does not mean they are not true or real.

The shaman told me that Orson had come to see me, that he had lain in his familiar position with his head on my right foot while I sat and wrote at the computer.

I had not told her that I had felt recurring pressure on my right foot in the weeks since Orson had been euthanized.

She also said that Orson had come to see

Winston the rooster, his friend and companion. He had come to tell Winston that when he was ready, he would take him to the other side, to the world of the blue lights by the stream.

She did not know that Orson and Winston were friends, that they napped together in the yard, that Winston was the only thing that could calm the excitable dog and make him tranquil.

She did not know that when I buried Orson up on the top of the hill behind the farmhouse, Winston the rooster, then old and lame from a hawk attack, came hobbling up the steep hill to sit by his grave. Winston would indeed die a few months later, and I often think of these two unlikely pals, sitting by their stream, dipping their claws and paws in the water, planning their next chapter.

When the shaman left, I thought I would not hear from Orson again, but I have felt his presence many times. In the summer of 2013, while herding the sheep with my border collie Red, I was standing quietly, as I love to do when herding. Red was watching the sheep graze.

I felt the old pressure on my right foot. I recognized it right away. It was Orson visiting me, coming down from the place of blue lights, perhaps, or from the life of another human being whose journey he was inspiring or marking.

I felt his presence. I think Red felt it too. He looked away from the sheep — something he rarely does — and the ruff on his back went up. I held up my hand to tell him to stay and be still, and he did stay, but he kept looking at me, as if transfixed.

Once in a while — early in the morning, especially — when I am alone and writing in my study, and it is dark and peaceful, Red will growl softly and I will see the hair on his neck rise. I will feel the pressure on my right foot, and feel the air stir, and I think it is Orson, coming back to me, checking on me, passing through on his way to a place I will never see or know or fully understand.

One night, when I felt a presence in my office, I had a powerful and very clear image of Orson sitting amid four rings of fire. He was at the center. The four rings formed a box around him, and it seemed to me that they were guarding him; they were a sanctuary, the place from which he came and went out into the world, and then returned to rest and heal his wounds and replenish his spirit.

There, in those rings, Orson was a different spirit, a different creature. He was calm, at ease, at peace. He was no longer trying desperately to fit himself into a life he could not manage or understand.

I was awake, sitting in a chair. This was not a dream. Orson was looking directly at me. He spoke to me, very clearly, through feelings that came to me in words and ideas that I could grasp.

I came to awaken you, he said, *to light the spark that would help you find your destiny, your life, your love. This was what you were looking for, what you were seeking your whole life.*

Lucky did not live long enough to help you; Julius and Stanley were not powerful enough. You were not ready. You needed a stronger spirit to confront you and awaken you.

This is my home, the home of my ancestors. This is where I belong. Each of the rings means something different — one is hope, one is love, one is destiny, the other is chance.

I know you were disappointed in me, but I did what I had to do. I did what I came to do. I left when I could do no more.

That is the ongoing story of Orson and me, the story of hope, love, destiny, and chance. It brought me to the place of miracles and limits, and to the roots of a new understanding.

4
ELVIS

When I bought Bedlam Farm in 2003, I splurged on good fences with thick posts and charged wires. I wanted my new home to be secure for the animals I planned to bring there. I was soon subsumed by the intense rhythms and complexities of owning a farm — the management of hay, fences, water, gates, barns and barn doors, winter, flies, mice and rats, coyotes and foxes.

I had two large pastures, each about seven acres in size. The first, behind the farm-house, held the donkeys and the sheep. The second was empty. I had this idea it might be fun to have a couple of cows back there. I had never owned cows, but I saw a lot of them upstate and I liked their genial calm and their ravenous appetites. People with farms learn small economies. If I had a cow, I wouldn't need to brush-hog my outer pasture.

I discussed this idea with one or two

farmer friends. This was the pre-Internet, pre–social media way of communicating. In the country, it's still the best way for farmers to talk to one another; very few have Facebook pages or the time to post all day. They are almost exclusively allergic to computing. This is a culture that believes if you have something to say, you say it face-to-face. If you want or need something you wait until you run into somebody at Farm Supply or at the coffee counter of the convenience store. You tell somebody who tells somebody and eventually somebody else gets in touch; the phone will ring, somebody will pull over his or her truck and give you a name. I accept and embrace change, but when people stopped talking to one another, connection and community suffered.

I am grateful to live in a place where people still talk.

So, I shouldn't have been surprised when Peter Hanks, a friend, photographer, and dairy farmer, pulled up to the farm in his pickup. He got out, hemmed and hawed, then retreated into the safe realm of weather chat, the farmer's staple conversation opener.

Peter was uncharacteristically hesitant that day, even shy. He seemed almost to be embarrassed by what he had to say. Few farmers I've met are great talkers; they have a lot of emotion, but they don't share or show a lot of it. Finally, he came to the point. He had this huge Swiss steer, Brownie. "There's something different about Brownie," Peter said, scratching his head. "The way he looks at me, follows me up and down the gate. He's not like any other steer I've had," he said.

I understood Peter's discomfort. Farmers know the difference between animals and pets, and farmers who start getting soft about steers can end up feeding an animal the size of a pickup truck for a long and expensive time. Farmers love their cows, but they also know that animals are their livelihood. Profit margins on a farm are slim, things are uncertain and unpredictable, hard decisions often have to be made.

"It was time to put the steers on the truck for market," Peter said. "For the first time in my life, I just couldn't do it. I couldn't put Brownie on the truck."

He had heard that I might be looking for a cow. Here was a steer, close enough. He would find a cow for me to go with the steer. What about it?

I thought it was crazy, at first, and of course it was. But the idea of learning how to talk to animals and understand them was growing in me. Somebody was now offering me a three-thousand-pound Swiss steer, an animal I knew absolutely nothing about. I had aquired the farm to understand animals and write about them. This was a gift, I realized, a wonderful opportunity.

At the time, I didn't know whether my experiments with visualization and other forms of communicating were going to become a more serious part of my life. How far might I go with the idea? Were my experiences flukes, or had I stumbled onto something that was real?

There were two ways to find out — amp up my own experiments and ask other people to try with their animals. Over the next few years, I was to do both.

I decided to rename my new steer Elvis. He just looked like an Elvis to me.

Just like Peter said, Elvis was an unusual steer. He seemed drawn to humans; he made eye contact and liked to follow people along a fence line. He projected a gentle curiosity and intelligence that neither Peter nor I had ever seen before in an animal like that. Elvis was enormous, but seemed genial. He moved slowly, seemed calm, even meditative. When I came up to the pasture gate, he would lumber over and look me right in the eye, as if he was waiting for me to speak. I could see how Peter got attached to him.

Since I didn't want to have a steer who was all alone — herd animals never like to be alone — I took in a second steer, Harold, from a different farmer Peter knew, and a female beef cow, Luna, as well. Both animals needed homes. Elvis, Harold, and Luna all occupied the rear pasture.

Around noon every day, Elvis would come and stand at the gate and stare into my study, about fifteen yards across the road, his brown eyes wide and pleading. I'd look up from my work and swear that Elvis was staring right at me. I stared right back.

This unnerved me; it was not how bovines behaved. I decided Elvis must be hungry and went outside and brought him more hay. As the days passed, he showed up more regularly at the gate. "You're training me, aren't you?" I shouted to him as I tossed yet another square bale of hay over the fence.

I saw that the square bales I had been feeding my donkeys and sheep would not last long with Elvis around. I found a hay farmer who sold 1,200-pound round bales and he delivered a dozen. I had to buy a small tractor to move into the pasture the round bales of hay that Elvis, Harold, and Luna needed to eat. The three of them ate a whole giant bale — it stood three or four feet higher than me — each week.

Elvis was the largest animal I had ever been near. I began to wonder how I might control and communicate with this enormous creature. On a real farm he would have been kept in a gated stall where he could be fed and controlled. Here he was loose in a seven-acre pasture. I couldn't imagine how I might treat a wound, have him examined, or stop him if he decided to run over me or someone else.

It is, I know now, a profound mistake to take in a steer as a pet, the worst kind of emotionalizing of animals. Steers are raised for meat and are not bred to live long. They are genetically programmed to breed and die a year or two later. Any longer and their legs begin to go, unable to support the animal's enormous weight. When this happens, they can suffer terribly. I've written extensively on the dangers of emotionalizing animals and yet I projected all sorts of things onto Elvis — he was cute, he was sweet, he liked me. I think everyone who lives with animals does it. When does it become too much? When it takes us over, causes us to lose perspective, obscures our ability to understand the real needs of real animals.

Elvis buzzed through my seven-acre pasture in a matter of weeks, drooling, snorting, buzz-cutting the grass as if he were a giant sucking machine, which he was. Like most animals, he loved food, paid close attention to it. I decided to see if I could talk to him, listen to him, even train him. I had no way of making him do anything — coming to the barn, standing still for a medical exam or a shot or fly spray — that he didn't feel like doing. This was dangerous. I had to be careful and thoughtful. Elvis could knock me down with a swing of his enormous head or a swish of his tail. If he ran over me, I would not be getting up soon, if at all.

If Elvis had thought to walk right through my five-wire fence, he could have. For that matter, he could have knocked the barn down, walked right through the wall. I had nightmares about him walking through the fence and into the road, where he might well be hit by one of the speeding pickups that tore down my hill most days. I wasn't sure who would survive a crash like that, though I wouldn't bet on the truck.

If Elvis got belligerent or excited, he could easily have harmed me or anyone working on the farm. He was so big it took him ten yards just to stop when he was walking downhill, and when he decided to lie down, the pasture shook. He was surrounded by vast clouds of nasty flies, which he did not seem to notice, and he had a penchant for drooling and plucking and eating hats right off my head.

I was about to undertake a whole new level of communicating with animals. And I was scared. I imagined myself standing in the pasture conjuring up images for Elvis to receive while he simply walked right over me and trampled me into the ground. Much smaller steers than Elvis do that all the time to farmers who are cocky or not paying attention. My ideas would either work or they wouldn't, but I was determined to try. I sat down and wrote out a plan for talking to Elvis. I had never communicated with an animal the size of a small cottage; I decided to proceed slowly.

First, I got a huge basket of apples and put them inside the barn. I stood outside the pasture gate and held a few up. When Elvis saw them, he came trotting over, rattling the gate and shaking the earth like the giant chasing Jack down the beanstalk. I held an apple out in my hand but did not offer it to Elvis until he was still and calm.

Giving Elvis an apple was like giving a pea

to an elephant, but he loved them and chewed them deliberately and with gusto, drooling bits of apple and spittle all over the ground. After a week of this routine, Elvis began to pay closer attention to me. When he saw me come out of the farmhouse, he trotted to the gate and waited for me. I asked him for nothing, gave him no commands, and moved slowly and predictably in his presence. I was establishing trust.

Elvis was used to seeing humans in tractors dumping feed. There had been little one-on-one interaction with humans. This was as new to Elvis as it was to me.

Remembering that food means life to animals, I brought Elvis his apples three times a day and waited while he chewed each one. Sometimes, when I came out in the morning, he would be waiting for me, anticipating my arrival. Sometimes, I saw, he got excited, snorting and drooling, rushing toward the fence. If I had been inside the pasture, I might have been pushed right up against the gate.

But food is only a first step toward gaining trust, toward opening the doors of communication with animals. It can be a gateway to visualization but is very different from it. Food alone is training and trust; it is not really about understanding.

After three weeks, I opened the gate and came into the pasture. The danger here was Elvis getting excited and running into me. When he came toward me, or got too close, I held the apple behind my back, then walked out of the gate. After two or three days, Elvis learned to stop. He never got the apple unless I held up my hand and said in a soft voice, "Come here." He caught on quickly.

A few mornings later, after I was sure he grasped the idea of waiting, I decided to try visualization, the first test. Elvis leaned over me and tried to nudge me with his giant head; he wanted me to fork over the apple he knew was behind my back.

But I didn't have an apple.

I paused and held my hand up in a kind of "stop" gesture. If he got too close, I flicked him lightly in the nose with my thumb. He looked startled and froze. Flicking isn't a sophisticated or especially positive training method, but when you are training a three-thousand-pound steer, a flick is less troublesome than a fly bite. And it's gentler than the methods many farmers use.

I closed my eyes and cleared my head. Animals like Elvis can sense human feeling and emotion; they can literally smell it.

Expressions on my face, my body language, my demeanor and state of mind are all details that animals focus on, even if people miss them.

It was critical that I feel confident and look confident. It was important that I believed in what I was doing — animals read that. I stood up straight, puffed out my chest a bit. I moved away from Elvis and took a deep breath, telling myself that I was absolutely certain about my ability to control this great beast. I pictured it, pictured it again, made every effort to project it.

Elvis was drooling prodigiously now in anticipation of food. I took a deep breath, told myself again I could do this, and imagined him standing still, obeying my command to stop. If it didn't work, I would have a big disappointed steer coming rapidly toward me.

It did work. Elvis seemed to understand what I wanted. We both stood stock still for the longest time, perhaps three or four minutes. I projected images to him conveying calmness, stillness, patience.

He didn't move. This was exciting. I could feel that we had communicated in some way. The apples had established a bond — I had gotten his attention, which is important

— but I had moved past that bond, deepened it. I was building the ladder I would climb to reach a better understanding of animals. First, trust, then attention. Visualization, exchanging the images of life. Then projecting confidence and clarity. Then listening and, finally, communication.

I was able to project the right emotions — confidence and clarity — and Elvis was doing exactly what I had asked him to do in my mind. And I had received images back: Elvis chewing contentedly on his apple, Elvis standing still in the pasture alongside me. He still wanted the apple, but I had also gotten another image into his head, another idea. To be still and wait.

While many behaviorists choose to make animal lovers happy by reporting on how smart animals are, and how they are just like us in so many ways, Temple Grandin, a gifted animal scientist and author, has devoted much of her life to studying how different they are from us, how they really think. She has broken new ground in learning how to talk to animals and how to listen to them.

More than any behaviorist I have read or studied, Grandin has learned to see the world in much the same way cows and pigs and sheep see the world. And as someone with autism, she has also studied the profound ways in which autists and animals often think alike, thinking through images rather than words. Two of her landmark works, *Thinking in Pictures: My Life with Autism* and *Animals in Translation,* have influenced me greatly, affirmed many of my instincts, and kept me on the right path.

Growing up as an autistic child, Grandin writes, she learned to convert abstract ideas into pictures as a way to understand them. She visualized concepts such as peace or honesty with symbolic images. In her mind, peace was a dove, honesty the image of placing one's hand on a Bible in court.

I read dozens of behavioral journals each year, and sometimes I think that behaviorists are not immune to the sources of funding or the passions of the moment. No behaviorist ever got a fat grant for writing that animals are not as smart as we think. Every year I read another story about how many words some border collie knows — eight thousand at last count. I have lived with border collies for two decades, and though mine are very smart, they are not linguistic geniuses.

I don't really see the point of teaching them a lot of words. That isn't how they communicate with the sheep, one another, or with me. Like Grandin, I celebrate their differences from us. I don't need to make them a mirror of me.

They know their names, "lie down," and "get the sheep," and while they may know thousands of words, they don't know the difference between a plow truck with a diesel engine and a flock of sheep: they will try to herd both, given the chance.

In recent years, the idea that animals are just like us, with our thoughts, fears, jealousies, and loves, has become epidemic. As someone who is regularly invited to speak about dogs, I can report that the very idea that dogs and human children are different is a shocking and controversial concept for millions of people. However, my idea is that animals and people are very different, and our failure to understand that makes real communication with animals almost impossible.

In *Animals in Translation,* Grandin writes that "animals and autistic people don't see their ideas of things; they see the actual things themselves. We see the details that make up the world, while normal people blur all those details together in their general concept of the world." We can't rely on words to communicate with animals, no matter how many they may understand. We need to see the world in the same way they do, and that means seeing it visually and in great detail.

Animals see details people don't see. They can smell our emotions and read our body language and intentions. When I put a syringe or medicine in my back pocket and go out to the pasture, my donkeys will take off before I get near the gate. If I put a cookie in my jacket pocket, they will bray joyously at the sight of me and wait until I get to the gate. It happens every time. They can look at me, the way I walk and hold myself and look at them, to determine instantly if I am coming for business or pleasure. They listen to me, they see me, they understand me.

Animals have senses and perceptive powers beyond our imagination. Just as Grandin learned that a shadow on the ground or a sudden change in light can frighten a cow, so I learned that everything about me — what I am thinking, feeling, wearing, intending — is important to a dog (or a steer), is a fundamental part of the way we will communicate.

It is not helpful to think of animals as being just like us, or to celebrate the idea that they are more like us than we could possibly imagine. That is a falsehood that diminishes both species. It distracts us from the work we need to do to understand the real nature of animals. In many ways they are less evolved than we are. They do not have the power to improve their lives or have consciences; they seem helpless against the ravages man has inflicted on the earth and on their lives and habitats. In many ways they are better than we are — more accepting, less destructive, more intuitive.

I called Peter and told him I had trained Elvis to stay. I said I could get him to come on command as well. He didn't believe me. An hour later, he pulled into my driveway in his pickup with three other dairy farmer pals packed in.

All four grizzled, sunburned farmers piled out of the truck. They crossed their arms and stood doubtfully by the gate, waiting for me to enter the pasture and get a steer to stay.

I was anxious. I knew there would be consequences if I was wrong. Within an hour, the story would be all over the county that some dumb outlander from New York City had tried to get a Swiss steer to stay and gotten his ass spread all over the pasture. And to be honest, I wasn't one hundred percent sure yet myself. Visualization sounded a little strange, even to me. Had I really communicated with Elvis in this way? Really gotten him to stay? Or perhaps he was just confused. Maybe I was.

Peter took me aside. "Look, no offense," he said, "but I don't want to see you get hurt. You sure you know what you are doing here? I've been around cows my whole life, and I've never trusted a big steer like that — and I like him — to stand still just because I wanted him to."

I explained to Peter that it was more complicated than that. Just stand back and watch, I said. Peter said I reminded him of the guy from Siegfried & Roy whose trained tiger mauled him nearly to death. "He thought he could trust the tiger, too," he cautioned.

Well, it wasn't Las Vegas, but I had my own show to put on. I saw that Peter had brought along a cattle prod. When I asked him about it, he said he was going to give it to me so I would have some protection if things went badly. I thanked him, but declined. I opened the pasture gate and walked in. I was actually emboldened by my audience. I felt I had been given the chance to prove something — to them and to me.

The farmers were watching eagerly, like hockey fans sitting right behind the glass. The smirks on their faces were pronounced and visible even from afar. Elvis was standing about fifty feet from me. He turned and started to drool, his enormous brown eyes opened wide. He started to trot, and it was like a hill rolling toward me.

I took a deep breath, held up my arm, and said, "Stay." I said it in a loud, clear, and confident voice. As I spoke, I pictured him standing quietly in the pasture. He slowed, but kept coming. I pictured him coming to a dead stop, I cleared my head of words and other images and focused on that picture.

Elvis took a few more steps, then came to a halt.

I heard the gasps and exclamations from the gate — "Jesus Christ," "Look at that," "He got him to stay," "He did it." Elvis was drooling great quantities of spittle, but he was fifteen feet from me, and there were no apples. I waited two or three minutes, then I walked toward him at a steady pace.

I patted him on his neck, talked to him enthusiastically and in a high voice, then walked back to the gate and waved to him to come to me. Domesticated animals love and need attention. Elvis quickly attached to me. He got treats like apples and carrots and he relished the attention. He even got sprayed so that the horse flies stayed off him.

I was confident now, and focused, and I could actually feel the connection as if there was a channel of energy connecting the two of us; emotions and thoughts were passing back and forth between us.

I knew what I wanted Elvis to do and he knew what I wanted him to do. We were in sync, together. It was not a case of my giving him orders, or shouting commands at him; it was more symbiotic, a real partnership. He got to live instead of go to slaughter, the fate of most steers, and he got extra care and affection.

Elvis was not my child, or even my buddy. He was a fellow traveler on my journey through the world. We were different, yet we were connected through the experience of living. This is the spiritual part of communication.

I took out a syringe filled with antibiotics the vet had prescribed for an infection in one of Elvis's wounds, as well as some salve for his wounds and bites. I saw the farmers tense up again. When their animals needed treatment, they would confine the steers and cows in narrow and tight stalls where injections could be administered through metal bars and they would be protected against kicking and lurching.

I simply showed the syringe to Elvis, gave him a big carrot, imagined the needle going into his side, and stuck it in. I was careful once again to precisely visualize what I wished him to do, and to remind myself that it would happen, to be confident and sure.

Elvis snorted and flinched, but he did not move. He crunched his carrot and looked at me.

The farmers were impressed. They said they had never seen anyone get a steer to "stay" or "come" or be still for a big needle. They peppered me with a score of questions. Their tones had shifted from dubious to curious.

Their support was important to me. Farmers have taught me some of the most important lessons in my research about talking to animals; they are among the world's greatest animal lovers. They know their cows in the most intimate and personal ways; they see them every single day, they milk them, breed them, midwife their births, treat them when they are ill.

As part of my research, I spent months on farms and inside dairy barns, watching the ways in which farmers understand and communicate with their animals. I was shocked to see how one man could easily control eighty large dairy cows, often in the dark, and even in rain and snowstorms. Ed Anderson, a farmer in Jackson, New York, used a number of different techniques to control and communicate with his cows.

He would rattle the pasture gate to alert them to milking time, or utter a low whistle when grain was put out. The cows would appear out of the darkness and mist, come through the gate, cross the road, and each go into their own stall, where they would wait to be grained and hooked up to milking tubes.

Ed would talk to the cows and sing to them, but most of the time, he just tried to think like them. "We talk to each other all the time now," he told me. "They let me know when they are hungry, sick, or scared. I let them know when they need to stand still for some milking and be calm."

Ed could stop a herd from crossing the road simply by turning sideways and just coughing — it is a sound, he says, that startles and confuses the cows and causes them to stop. He could also stop them by putting a bucket in front of a doorway he didn't want them to enter. If a strange object was in the way, they would stop.

Ed says he is so close to his cows that the lines of communication are sometimes blurred. When he comes into the barn in a bad humor, the cows are quiet. When he is happy, there is a great din of "mooing and muttering," as he calls it. By now, he says, the communication is mostly wordless. I pressed him to describe how that communication happens, and he talked about receiving images from his cows. He introduced me to the idea of translating images. "It isn't always what you see," he said. "Sometimes it's an image you have to interpret. You have to think about it sometimes. You can't solve an animal's problem unless you put yourself in their place; then the images make sense."

For example, he once got an image of cows running in a panic down to the pasture gate. He got it every day for nearly a week. He got on his ATV and drove up to the top of the pasture hill, and looked through his binoculars and saw a mountain lion running through the woods across the valley. He kept the cows in the barn for a week, and when he let them out, there was no sign of the mountain lion, no more images. But he was certain he had received a message.

I have interpreted many such messages from my animals, especially the dogs and the donkeys. Once I saw an image of the donkeys lying down, and I checked their legs and found an infected scrape on the knee of Simon, my oldest one. I got an image once of my Lab Lenore rushing frantically out in the backyard. I took her outside, and she ran into a corner and vomited. I received a message in the form of an image from one of my border collies — a quiet place, almost a nest, under some rocks — and I knew she was very ill, something confirmed by my vet.

Word of my experiments with Elvis swept quickly through the farming community in my area, and soon I was giving demonstrations and discussing my theories. Ed Anderson was one of the first farmers to come by and watch Elvis. Elvis stood and came for me, and stayed still for a few minutes for good measure. As a final gesture, he came over to the gate, pulled Ed's hat right off his head, and ate it. He did not tell me he was going to do that.

Elvis and I had a sweet relationship over the two years that I had him. He nearly bankrupted me with the hay he ate and the equipment I had to buy to maintain his lifestyle. As expected, his great legs began to give out. The vet came to look at Elvis's legs and warned me that they would soon be unable to bear the steer's weight. He would fall to the ground, he said, and be unable to get up. His great weight would cause his lungs to collapse; he would struggle to breathe and either die a painful death or have to be shot, assuming someone with a rifle was around to notice his plight. The vet urged me to euthanize Elvis before he suffered further.

Another factor was that I could no longer afford to keep Elvis, and it was not rational or appropriate for me to do so. Steers are not pets, and ought not be confused with pets. It was one of the most painful lessons I learned in my search to understand the true nature of animals. It was also one of the most important. A farm teaches you when to let go, and when not to let go. I was learning when to let go.

Reluctantly, I arranged for Elvis to be taken to a slaughterhouse known for its quick and humane methods. When they came for Elvis, the truck pulled up to the gate, then two men got out and put up metal barricades on either side of the back door. They were nervous when they saw the size of Elvis — he had come down to the pasture when he saw me, and he stopped and stared at the strange truck.

The men shook some grain, tossed it on the ground, and made a trail into the truck. It usually works, one said, but Elvis wasn't buying it. He came lumbering over to me, drooling, checking my hands, waiting for instructions. I felt a powerful stab of guilt and betrayal; Elvis trusted me. I looked him in the eye and told him the truth, speaking the wordless language we shared. "I won't lie to you," I said. "This is the end of our time together; you won't be able to walk much longer. You are going to get on the truck you've managed to avoid all these years." I closed my eyes, cleared my head, and showed him where he was going. I broke up a bit, and felt tears starting.

I waved to him to come forward, took an apple from my pocket, and tossed it into the truck. He ignored it, walking up to me instead. I turned and walked up the ramp into the truck — the two handlers were standing outside the pasture gate, wide-eyed at the sight of me walking into the truck with a giant steer.

Elvis followed me. He stood calmly and patiently in the truck as if he did it every day. He ignored the grain and the apple. I leaned over, tears streaming down my face, and kissed him on his great big nose. I walked down the ramp, and I could not look back. The two men came running forward and quickly closed the back door of the truck In the end, Elvis would feed homeless people at a shelter in Glens Falls, New York, for an entire winter.

It tore me up to put Elvis on that truck, but I do not regret it or apologize for it. Anyone who has lived a life with animals or has been on a farm will understand it, although I well know that some will never understand or accept it. I am grateful to Elvis for many things, and one was his grace when it was time for him to leave. I believe he understood what I was asking and knew where he was going. Animals embrace acceptance in a way that is unnatural and alien to human beings.

By the time he left my farm, Elvis and I were communicating daily. He would stand, come, and be still for me. I could give him a shot, or rub balms and ointments on his scratches and sores. I knew when he was hungry and when he was in the mood to sit quietly with me on the hill and look out over the valley — this was an image he communicated to me often and clearly, perhaps more clearly than any other.

Elvis took my ideas about talking to animals to another level of understanding and skill. If I could talk to a three-thousand-pound steer, I could talk to almost any kind of animal I might encounter. I felt Elvis and I had broken through. I think we genuinely saw and heard one another. It was not, for me, that complex or mysterious a process. It required me to look at animals in a new and different way. It required me to accept that, for all our yearning and breathless discoveries, animals do not speak our language, and we do not speak theirs.

We can work as hard as our fevered imaginations would like, but we cannot turn them into us, nor us into them. They are different, they live in an alien world. Our task is to look for the windows and cracks where our souls and spirits meet, to move bravely and lovingly through the openings they make.

5
ROSE'S MESSAGE

I bought Bedlam Farm to expand my life in the country and my life with animals. I had left my family, friends, and work behind to live as different a kind of life as was possible for me. I knew I was unprepared for the rigors of farm life and bought Rose, a border collie puppy bred from strong herding lines, to come to Upstate New York with me and help.

Despite the painful end to the Orson story, I was excited to have another border collie. I was mesmerized by the energy and intelligence of the breed, and as a fellow ADD, I connected with him on a spiritual level. I had learned a lot, much of it the hard way, and I welcomed the chance to apply it.

When we arrived, Rose was six months old. I knew next to nothing about farms, and I scrambled to do the obvious — fix up the barn, patch the holes in the roof, get good fencing.

In October, a trailerload of sheep (and one donkey) arrived from Pennsylvania. I had just had more than a year of sheepherding lessons, and the plan was to wait until Rose was a little older and then train her slowly and gradually.

When winter hit early that year, I could not possibly have been less prepared. A blizzard from Canada came roaring down with shrieking winds, covering the farm in snow and ice. The farmers called it a Canadian howler, and for good reason. The wind sounded like dark spirits loosed from the grave.

The sheep and donkey had never been in such a storm. They panicked and tore out through the pasture gate, then ran across the road, down into the woods, and out of sight.

I had no idea what to do, and there was no one around for me to call. I stood there helplessly. The wind and cold and snow were brutal, beyond anything I had experienced. And I had a six-month-old puppy with me.

Rose was not panicked by the storm, even though her fur was quickly covered in snow and ice. She did not blink or flinch or take her eyes off me. Neither of us had ever been in this kind of a situation. How would I get those animals back to the farm and through those gates and up into the Pole Barn?

Six months is too young to seriously train a border collie. I could not possibly start her training by letting her loose in a blizzard and have her wandering around the woods looking for a donkey and the sheep.

I stood in a blue hooded sweatshirt, no gloves, and city shoes. I was covered in snow and so was my puppy, who looked out toward the woods where the animals had gone and then back at me, pleadingly. Rose lacked any commands to follow, or any experience with these woods. I could not imagine how the animals could be made to come back to the farm, even if we did find them, and they could be miles away by now.

The puppy was not on a leash, but already smart and intuitive enough to know not to run out there without my permission. I held my hand up, like a police officer stopping traffic. If Rose could speak, the words would have been "send me in, coach."

And so, we had a conversation.

"You're too young," I said. "If I release you, then I'll have my dog and all of the other animals out in the storm."

Rose: "I can do it, give me a chance to succeed."

I looked at her, and around at the blizzard, almost a whiteout. The wind was howling down the hill; the snow was already drifting. I thought of Rose out there in the woods, the animals stuck out in the open during an awful storm that was expected to last a couple of days and get much colder. If I waited much longer, who knows how far the animals might get?

I knew it was dangerous to send Rose out there, but I sensed her strength. She was ready, eager; she knew what she was doing.

Sometimes you have to follow your heart and gut, not your mind. "Go get 'em, Rosie," I said. "Find them for me." It was the first time I gave Rose that command; it would not be the last.

Rose needed no more encouragement. She took off across the road and through the big meadow and vanished into the woods. I ran into the farmhouse, put on some boots and a heavier jacket, and ran out after her. I could only see a few yards ahead of me, and the wind was deafening, driving the snow into my face.

I was already berating myself for this dumb and irresponsible decision. Rose was untrained and young. What if she got lost out there as well? Could she make her way back? Would she keep running, or get picked off by one of the coyotes I had heard yowling at night? How would she find the animals in this weather?

I saw her tracks for a few yards, then they disappeared in the snow. I ran to the end of the meadow, about two hundred yards from the farmhouse. I called out to her. For five or six minutes, I heard nothing.

Then I heard barking. The sound was sharp and clear. I felt it was directed at me. It was insistent, as if it were an aural locator beacon that I could follow. Rose was calling out to me. She had found the animals and was letting me know where she was. I followed the barking, regular and continuous, and came out of the woods and into a long culvert by an abandoned dirt road.

I came closer and saw the donkey and sheep, all huddled in the culvert, Rose in front of them, charging and nipping at any one of them that moved. Scientists say that dogs have vast stores of genetic memory; their instincts call up images from the past like files in a computer, only faster. Rose instinctively knew how to herd, given the right training and exposure, and at the right time. It is easy enough to mess up herding dogs by pushing them too hard and too soon.

Rose had followed the animal tracks

through the snow, or smelled the donkey and the sheep, or heard them walking through the brush and snow. The sheep instinctively drew into a flock, as they do when a herding dog appears, and the donkey, also a herd animal, stayed with them.

Rose was barking and racing back and forth, keeping the sheep together, creating her own portable border. I was astonished to see this puppy doing precisely what I had hoped she would do and wanted her to do. Rose and I had not had the chance to do any training together; I had no working commands to give her yet. I was communicating with my emotions — concern, focus — and my mind, imagining a walk back through the meadow.

And that is what happened. Wordlessly, I took a deep breath and told myself that this would happen, it had to happen. I imagined it and remained calm and clear. I did not throw a lot of words at the young dog, or project anxiety or doubt. I moved behind the culvert. There was nothing between Rose and the pasture gate but a meadow and a blinding, raging snowstorm.

I hoped she would sense my intentions and feelings. I hoped her instincts would pull up all the genetic herding memory that had been passed on to her.

As I moved farther back, Rose ran over to me, and the sheep made good on their opportunity to move out of the culvert and away from the large man and the barking dog. I spread my arms wide and moved forward and the donkey — ever willful but anxious, it seemed to me — moved out behind them.

I think these animals were looking for their shelter and for some hay and water. I held out my hand to Rose — a silent stay command. She saw it and grasped it. She stayed put, giving the animals a chance to follow their own steps back. And this was a tricky part. If Rose moved too fast or panicked the sheep or the donkey, they might bolt into the woods and go even farther away, or worse, run down the road and into the town. There would be trucks and plows down there.

It was not a good place for any of us to be. I ought to be in the farmhouse with the dog; the sheep and donkeys ought to be up in the Pole Barn, dry and protected from the wind, snow, and ice.

Rose marched the animals through the snow, across the road. She was firmly in command, rushing back, forth, and sideways to keep pressure on them all, to keep them moving forward.

And they seemed to be grateful. As they came within sight of the big barn, the animals rushed forward, picked up speed. They wanted to be safe and warm and dry again. They wanted to have hay and fill up their bellies on this savage night.

As they neared the road, I looked up in alarm. I saw the flashing lights of a town plow truck coming up the hill. Rose could barely see me in the snow. I felt a wave of concern. I remember creating an image in my mind of Rose running up around the rushing animals and heading them off at the road, holding them until the plow truck came by and I could get there. I was already running.

I remember feeling confident about this image, confident of Rose. And I remember the sensation of sending this thought out of my head, across the meadow, and of Rose receiving it. I was not fearful any longer; I knew she was getting the message. I could almost see it traveling through the snow and the wind. It was a powerful moment for me. I saw and felt my message getting to her, saw her reacting to it. It was all so fast, so unconscious; it had nothing to do with commands, and everything to do with trust, connection, feelings, and images.

Rose stopped weaving. She tripled her

speed and burst through the snow. She got out ahead of the animals and stood her ground, giving them the border collie eye and barking. The sheep stopped, confused, and obeyed this little dervish racing back and forth in front of them, daring them to move.

I came up huffing and puffing. Rose gave me what I thought was a contemptuous glance — *about time* — and I got around behind her, saw the plow truck flash by honking its horn. I got out into the road, saw it was clear, and told her to come ahead ("come on, Rosie"). She circled around to the other side.

I opened the gate and the animals came thundering back in. They rushed up to the shelter of the Pole Barn, Rose in steady pursuit. I closed the gate, got some bungee cords, and tied it closed tightly. I went into the barn and opened the wide doors, dragged some hay out, and put it into the feeder. I saw that the water tanks were full, and called Rose to me.

We went back into the farmhouse. I saw her shake herself off and go lie down by the woodstove. I sat down with a cup of tea. I was shaking, too, and not just from the cold. It was the excitement of realizing that I had been the fortunate beneficiary of a gift, that

I could talk to this dog. I was grateful for it, and I worked hard to develop it. It seemed lucky, not the fruit of genius or special intelligence. I had won her trust and attention, showed her my feelings and intent, shared my mind and needs with her. There was no longer any question in my mind that communication was possible; it had only to be refined and polished.

I was talking to animals.

The following spring, in the middle of lambing season, I was anxious. It was my first experience of lambing. Friends told me what equipment to get, but I was all alone on a big farm with twenty-five pregnant ewes. One night, I was sound asleep when I felt something nip at my arm.

Rose, who often looked out of the upstairs windows into the pasture at night, was nipping at my arm, which was hanging off the bed. She was trying to wake me up. She nipped at my sleeve and forearm, and then barked.

I was annoyed. What was she doing? My first thought was that I had to keep her downstairs, or retrain her. My instinct was to yell at her in order to get her to stop barking and nipping in the middle of the night.

What a nightmare this could be, if it happened often, if it got to be a habit. I thought of all the complicated dog training books I had read, how to discourage behaviors by pinching, crating, flicking fingers, using a sharp and corrective voice.

I sat up, and Rose ran out to the window in the hallway. I got up and followed her. I thought I heard a different kind of bleating out in the pasture, odd for the middle of the night. It was sharp and high-pitched, not a sound that had ever come out of any of my ewes.

And then I got it. I saw the look in my dog's eyes, the urgency in her bark, the excitement. Rose was the guardian of the sheep, their herder and protector. She always kept an eye on them, listened for them. She sensed coyotes before they did, knew when a sheep got stuck on a fence or had fallen ill.

A lamb! I wasn't expecting the lambs to be born for another month, but Rose had heard a lamb. It was a cold night, icy and windy, a dangerous night for a newborn lamb. I couldn't see the baby animal, but I could hear it. I got dressed quickly, grabbed a jacket and a big flashlight, and rushed downstairs and outside.

In the darkness, I couldn't see where the lamb was, but I opened the gate and Rose rushed up the hill and behind the Pole Barn. I followed, carrying the first aid kit I had prepared.

When I reached Rose, I shined the light and saw a newborn lamb, covered in amniotic fluids, wet and shivering. She looked as if she was nearly frozen to death. I took out a blue sling I had purchased and got it under the lamb so I could pick her up. Then I saw that the mother was struggling; she was lying on her side nearby.

I pointed to the ewe. "Get her up, Rosie, Get her up!" Rose looked at me hesitantly; this was not a command I had taught or used. I stopped, took a breath, and visualized Rose getting the mother up. Rose responded immediately. She came around, started barking, got the ewe on her feet, and then got behind her.

We were in sync, talking to one another now. Rose didn't know where to bring her. I sent her an image of the barn, where the lambing pens and heat lamps and hay and straw were. Holding the sling in front of me so the ewe could smell the lamb — this is how sheep bond with their babies, by smell and by cleaning them — I began walking slowly backward down the hill and toward the lambing pens in the barn, which I had spent weeks preparing.

Step by step, we came down the hill. Several times, the mother panicked, got confused, or tried to run. Rose headed her off, kept her moving behind us, pressured her so that she wanted to be closer to me and the lamb than the Pole Barn and the other sheep. This was critical, because if she ran away, then she might lose her connection to her lamb, and the lamb could die or would have to be bottle-fed for months.

I think of that march down the hill as a long conversation, just as clearly as if it were spoken in words. When I needed to go faster or slower, Rose knew it, sensed it, felt it. When the ewe got nervous, Rose saw it and told me, and I slowed down or waited for her to catch up. When we got down to the barn, Rose stopped outside and waited, so that the mother could come into the barn, and then Rose blocked the door. We had them both, we had done it. We had saved the lamb together.

I put the lamb on a pile of hay, turned on the heat lamps, opened the gate, and stood back. The mother looked around. Rose came into the barn quietly and the mother looked at her lamb and rushed into the pen. I threw in a tub of molasses mixed with water, cut the lamb's umbilical cord, cleaned the fluid out of her nose and mouth, and rubbed a soft towel on her to warm her up. Then the ewe, recognizing her baby, began licking and cleaning and nuzzling her. The baby found her nipple and began to feed hungrily. They were both all right.

For the rest of that spring, whenever Rosie barked in the night, I got up and we found the lamb and mother, and got them into the barn. Only one lamb died; she seemed to be sick at birth. All of the others made it and grew up to be healthy.

Rose had learned how to talk to me. She would come to the bed and bark, and if I didn't wake up, I'd get a nip on the arm. I had learned how to listen to her.

There was a reason why Rose's nips on my arm startled, even terrified me, why her messages came to take on far greater importance even than the life of a lamb.

Those nights brought me back to my very brief time with Lucky. I had dark troubles in my childhood stemming from one traumatic night. I don't remember much more than someone coming into my totally darkened room on the upper floor of our tiny wood-frame house on the east side of Providence. I heard him coming up the stairs, one creak at a time. My parents were not at home that night. I remember him opening the door as I felt a powerful terror unlike any I had experienced. When my parents came home, they were shocked to find my door open, and to find me crying and hiding under the sheets in a urine-soaked bed.

The police tried to get me to draw pictures of the man who came into the room, but I either didn't see him clearly or terror blocked my memory. I remember his breathing.

Various doctors tried different therapies and my father lectured me every night about choice and determination, but there was only one time in my early life that I did not wet my bed after that awful night, and that was the time Lucky came home with me and slept in my bed.

Lucky and I shared the first dialogue I ever had with an animal, even though I could not have understood it as such at the time. I believe he came to me to help me begin the long and arduous task of learning to trust, to be open and safe.

Every dog or animal I have had has been a magical helper on that journey. Lucky was the first. I sometimes think Rose was his spirit reincarnated, that she came to finish the work he began. I believe in spirit dogs. I believe they come when you need them and leave when their work is done, or when your needs change. Lucky was the first living thing to understand me, to sense my pain and fear, to help me climb over it.

In the six weeks he was with me, I never wet my bed once, and when he was taken away, I wet the bed every night until I was in my teens. Bed wetting shaped much of my early life, and so did the accidents I had in school, to the rage of my teachers and the ridicule of my classmates. I know now that this affliction is common in abused children, a powerful symptom of early trauma and abuse. I did not know then, nor did my father, who told me I was weak and lazy, that I could stop anytime I wished, if only I mustered the will.

I never answered him during those nightly lectures, but the truth is, I would have cut off an arm to stop wetting the bed. I could never go on a sleepover or have one in my home. I only went to camp once in my life, and never dared to do it again. It was a dreadful nightmare.

Lying in my bed in my dark farmhouse, I began to make the connection between Lucky and Rose, between my need to communicate with animals and the powerful messages they had been bringing to me.

I had come to my farm on a hero journey to find out who I am, to heal the wounds in my life, some inflicted on me, some I inflicted on others. Joseph Campbell writes that the hero journey is about leaving familiar places and heading out into the unknown, where there can be great danger and confusion. If you are fortunate, he writes, magical helpers, often in the form of animals, will appear to guide you and help you feel strong when you are afraid.

I came to see that Rose helped me through those first difficult years on my farm in many more ways than I had imagined.

At the farm, on a remote hillside, I often woke up in a sweat, in a panic. Freud believed that when we suffer a traumatic experience, we often replicate it in order to try to solve it. I was hundreds of miles from family or friends or any of the things I was used to. Back home, I was never more than a few feet from people, from help.

On the farm, it was a different experience.

In my years of therapy, I asked the same question over and over again, the question every traumatized child asks: "Where was everyone? Why did no one see or hear? Why did no one come? What did I do to deserve this, to bring this great shame on myself?"

I never got an answer.

I remember one night on the farm, I woke up in a sweat, my heart racing in an awful panic. I was nearly paralyzed with fear, the room was dark. I thought — as I had so many times — that someone was coming up the stairs. I was, as usual, alone.

I often have dreams, and they are rarely good ones, and this night was no different. I thought I heard Rose talking to me, telling me it was okay, I was not alone anymore, I was not a child any longer, I was not helpless anymore.

"It's all right," I felt her saying. "I came to tell you that you are all right, you will live your life, you are safe here. You can let the fear go; it's just a space to cross."

I had this powerful sensation of light and connection, as if something strong but gentle had burst through the wall of fear that had surrounded my nights for so many years.

I thought of Lucky, sleeping in a box on the floor, of how I picked him up and put him on the bed next to me. And then I thought of his rheumy eyes, his shaking and trembling, his saying an unmistakable good-bye to me.

I opened my eyes. I heard the wind outside, but it was different. There was Rose, standing by my bedside. I started to get up, confused and dazed. Was it a lamb? A coyote? A high wind? All things Rose might wake me up for.

But it was none of those things. It was something different. Rose was not nipping at me, not rushing to the window, not trying to get me up and on my feet. She was not warning me of some problem or danger in the pasture.

She had come to speak with me, to give me a message. I could feel it. She stood staring at me with her dark green eyes. I knew — felt — that she had sensed my fear, and responded to it. Perhaps this was why she had come; perhaps she really was Lucky returning to me.

Maybe all of the dogs in our lives are one and the same dog, coming back as long as they are needed, going home to change bodies and refresh and be healed and made strong again, for their strenuous and complex task of changing a human being.

I am here, Rose said. *You are not alone. You can figure it out. Talk to the little boy. Tell him it is okay, that he is not alone anymore.*

It was the most important conversation I had had up to this point with an animal. It touched the deepest parts of me and brought them to life. It integrated the adult with the child, and helped to narrow the gap between them.

I thought back to Lucky, and his purpose.

In my life with animals, I have learned they have messages for us, communicated to us in different ways if we are open to them and willing to work to understand them. Sometimes their messages are symbolic: Lucky telling me I was strong and could succeed. Sometimes they are literal: Rose telling me lambs had been born.

Lucky came to ease the loneliness and fear of a troubled child. His message was strength and safety.

Julius and Stanley came to help me with the sometimes frightening transition from corporate life to a creative life.

Orson came to tell me to move my unhappy life forward and enter into a compact with nature and the animal world.

Rose came to tell me I was strong and could survive alone on a farm.

All of these creatures had things to tell me. Each one guided me to a deeper understanding of animals and helped me understand how to communicate with them.

Rose was to save my life more than once. We talked to one another so often and in so many different ways that it became second nature to me. When we went out to work in the mornings, I simply pictured the tasks we were there to perform — moving the sheep, taking them out to the pasture, bringing them into the barn, holding them in place for the vet.

When I was anxious about a ram, Rose would get in front of him and back him off. When I lost a cell phone, I closed my eyes and pictured it and Rose went and found it.

Rose filled so many crucial roles in my life that had previously been missing or that hadn't been fully realized by those who were supposed to play them.

She was my mother — she was devoted to me, loved me, offered her life to protect me.

She was my partner — she shared the joy and travail of the farm.

She was my teacher — she helped me come to a wiser understanding of animals, far beyond the experience of having a pet.

She was my spiritual guide. The wolf lived and was present in Rose. She would turn into something wild and powerful in an instant; she connected to the animal world. And that was the mystical part, she was, in some ways, unreachable, beyond my comprehension. There were parts of her I could not talk to, could never understand, that she would never show me.

I loved Bedlam Farm. I wrote eight books there; it was a beautiful and inspiring place for me. It was the place where Rose and I lived together, worked together, learned together. It took a physical toll on both of us, though; the winters were brutal, the labor hard and unrelenting, the pastures steep and challenging.

Rose took more of a beating than I did. She was butted by rams, run over by ewes, raked by hidden barbed wire. Her paws were shredded on rocks, sliced by glass and nails. Every day was hard for her in different ways, and she had only one speed: fast.

One morning, I woke up and she was not lying by my side. She was not patrolling the hallway outside my bedroom. She was not looking out the window to check on her sheep.

I heard her call out to me, "Help me." The message was clear and sharp. I got up, grabbed my robe, and ran downstairs. I knew she was at the back door. It was the only place she would go if she was struggling and could not get to me.

I found her lying there in a pool of vomit. She was trembling and breathing heavily; spittle ran down her chin. She had had the first accident of her life with me, a puddle of urine on the floor.

"Oh, Rose, Rose," I said, dropping to my knees. I held her, leaned over her, whispered in her ear that I was there, that I would help her. Her sharp and clear eyes were clouded and glazed; she seemed disoriented.

She turned and saw me, focused on me. I felt her tail wag.

She looked me in the eye. "Help me to leave," she said. "Don't let me be like this."

Rose was the strongest, most independent, and least needy dog I ever knew. She was proud and never backed down from any duty, confrontation, or fight. She never curled up in my lap or whined for attention. She was at her happiest when she was outside with the sheep, watching them, moving them, being with them.

I had no confusion about what she was telling me, no doubt and little grief. Rose was in charge of her own destiny. She had come to help me, to stand with me as I learned to heal my fears. She did not wish to live an impaired or piteous life, or to be seen in that way.

I cleaned her up, picked her up, and placed her in the car, wrapped in a blanket. We drove to the veterinary hospital. Suzanne, our vet, examined Rose and shook her head. She thought there was severe neural damage, perhaps a stroke or a burst blood vessel.

Rose had struck her head or been struck in the head many times in our years together. I could, if I wished, take her to a veterinary neurologist in Massachusetts, Suzanne said. They could do an MRI; perhaps there was a tumor.

But she knew and I knew that I would never do that. Rose would not have wished for me to do that. I do not believe that the measure of love for an animal is to keep them alive at all costs.

Instead, Rose and I had one of our most important and meaningful conversations. It was time, she said, for her to leave the world, and bring her work elsewhere. Thanks to her, I was stronger; the nights had lost their dread for me. I was confident about running the farm. I knew how to give shots to the animals, when to call the vet, what kind of fences I need, what kind of water lines to run to the pasture, how much hay to buy and who to buy it from.

I had spiritually collapsed on the farm, nearly broken down completely in my confusion and aloneness. Rose brought clarity to my life, and gave me some strength. I do not know if I could have survived without her. I had gotten divorced and found love again all during my time with Rose.

Suzanne and I reached the same conclusion. Rose was deteriorating before our eyes.

I was determined to help her keep her dignity.

When I told Maria what I wanted to do, she burst into tears and challenged me. "No, we can't put her down!" she pleaded. She was angry, I could see it.

And then I cried, too, perhaps deeper and longer than I ever had before. I told Maria what Rose had said to me, lying by the back door that morning. She was asking me for my help, as I had asked her for hers so many times.

Please, please, help me to go, let me go.

Maria understood; she agreed. We put Rose down the next day.

I will always talk with Rose. Talking to animals is a mystical and ethereal thing. Once you open up the channels of communication with an animal, you become part of a rich and fulfilling conversation that can continue for all time.

Rose enters my consciousness all the time on the farm, when I am herding sheep, doing chores, walking up a steep hill. She came to visit me when I was in the hospital recovering from open heart surgery and she stands by my side whenever I wonder how to deal with all the things one as to deal with on a farm.

I think of Rose often, every time my border collie Red goes out to organize the sheep. My wish for her is the same, that she is running in a golden field, grass stretching to the horizon, sheep everywhere who need to be gathered up, led to pasture, taken home to the safety of their barns.

Be strong, she tells me. *Be strong.*

6
THE GREAT COLLAPSE

A friend who works in the financial industry called one afternoon in September of 2008. He was watching CNN. The economy was crashing; he had lost all of his savings in one afternoon. The world has changed, he said.

I walked out the back door of the farmhouse. I looked at the Kubota tractor I had purchased to move the giant, round bales of hay I had to buy for my two steers and dairy cows. I saw the sheep on the hill. I saw the cows in the back pasture. I saw my border collie Rose, who was still with us at the time, though lately she had been struggling to keep up with the sheep, her legs trembling and sore after years of hard work. She looked at me in a way she had not looked at me in all her years at my side. This is over, she said to me; this is something different.

Looking back, I ought not to have been surprised by the Great Collapse. I think

Rose saw it coming; perhaps that's why she decided to leave me so long before her time. She was, I think, trying to send me a message. Animals have instincts we do not have; they sense things we do not know. I doubt she knew what a mortgage or stock market was, but I am sure she sensed a great change coming in my life, in the world around her. Many memories are fuzzy in my life, but not the fall of 2008. Within a few weeks, my first wife and I decided to end our thirty-five-year marriage. This, coupled with those rapidly dropping numbers in the stock market, was the beginning of a breakthrough for me. These events were to help me see the world anew, to understand animals in a different way.

Overnight, I became a terrified five-year-old with a ninety-acre farm, a lot of animals, and a book contract. I was living a life I could no longer afford, did not understand, and could not maintain. And I was very much alone.

The Great Collapse became a bridge between one life and another, one way of understanding animals and one another.

I thought by that time — I had been on my farm for nearly four years — that I had learned a great deal about animals and how to understand and communicate with them.

Indeed, I had learned a lot. I was listening to the animals and they were changing my life. But it was nothing compared to what was to come.

We often project our needs onto animals, and when the Great Collapse came, I had two animals in the house — Rose, who ran the farm with me, and Lenore, a loving black Lab who helped me feel less lonely. I sang to her every night before I went to sleep. I didn't see until much later how sad a scene that was.

There were other animals on the farm then, too many. I had two Swiss steers, Elvis and Harold, a beef cow named Luna, three goats, six chickens, thirty-six Tunis sheep, and four donkeys.

Was I an animal hoarder? I certainly didn't fit the hoarder profile: the house was clean and in good condition, I had plenty of money (or did before the collapse), lots of pasture, good hay, and a pump that drew fresh water. I could care for them well.

But there was something wrong with my relationship to the animals around me. I had lost perspective. I was using them to fill holes in my life. I had too many animals to know them all well, to really understand and communicate with them.

And in a sense, they had helped me to lose

control of my life. I had gone a little mad up there on that hill, in those winters, continually hauling hay and water and calling expensive vets. I had been so insulated in the world of the farm that I could no longer see my own life clearly.

When I came back from a book tour in 2008, Anne, my bookkeeper, was waiting for me in the study. She had an awfully gloomy look in her eye.

"It's time to panic," she said after a long look at the bills and the books. I had run out of money. There was very little on the way. I didn't need to be told twice. I panicked and ended up sitting in a therapist's office in Saratoga Springs, New York. I have lost my life, I said. I want it back.

They say getting divorced is like being in a car crash every day for a year. It's an awful thing to have to separate from a good person who did nothing wrong, but our marriage just didn't work any longer. She loved her life in New York City; I loved my life in the country. We could not find a space in the middle, not even after so many long years together.

Like many people who get divorced and deal with feelings of guilt, I wanted her to be secure and I wanted the conflict to be over. So I gave away most of the money I had to end it. I didn't regret it then, and I don't now. It was the right thing to do.

The recession and the resulting collapse of publishing as I had known it took the rest of the money. No more big advances, no more regular royalty checks.

That was not the end of it. The recession had collapsed the real estate market as well as the stock market. I put Bedlam Farm up for sale, expecting to recover some of my losses, to build a new nest egg.

No one even looked at the property for nearly two years, and it was another two years before anyone bought it, and then only for a third of the original asking price. The farm wiped out the meager savings I had left.

I broke down, is the simplest way to put it. A nervous breakdown. The animals were swept up in all of it, of course. I let things get out of control. I lost perspective. I was not well.

I had purchased my three goats on impulse. I thought they seemed cute and bright. They were loud and restless. I did not need goats.

I had a helper on the farm who intensely emotionalized all of the animals there. Every day, each one of my thirty-six sheep got its own individual grain and food bowl — my helper didn't like fights.

The cows and steers were sprayed and brushed daily, sores and ointments applied to their bites and scratches. The goats got grain and cookies and treats all day long, so they bleated every time they saw any human being, and never stopped bleating. Each day, the chickens were fed special treats and mash by hand; they never stopped following me around. The cows mooed and bellowed at the gate; they associated people with an endless rain of treats.

I was oblivious to all this for a long time. I was deteriorating. I was heading for the Great Collapse. Exhausted and depressed, I had turned over much of the animal care to my helper, and I found myself with a life with animals that I did not want or believe in. I was disconnected from my own life and had fallen away from a true understanding of animals. I was not communicating with them, and they were not communicating with me.

One night during this period, I stepped out of the farmhouse to attend to some chore. It was bitter cold; the wind was whipping down from Canada. The ground was icy and I slipped on the ice and fell down, hitting the back of my head. I knocked myself out. I might have died out there in the cold, unconscious and alone. There was no one around for miles, no one to come by the farm and look for me, no one to call and check on me.

But there was Rose. I felt a sharp nipping at my ear. I heard barking and whining. *Get up,* I heard. *Get up.* I did. I struggled to my feet, and walked into a new life.

I saw that as much as I loved animals, they were not a substitute for human companionship, for love, for connection and community. I understood I was hiding behind them, using them as a substitute for the love I wanted, for the human being I wanted to lie next to in my bed, and to share my life with.

I knew what I needed to do. I told my helper that things had to change on the farm, that I could no longer afford these animals and did not want to have animals that were surrogate children. She quit in a huff.

This was during the time when Elvis, my steer, started to lose the use of his legs. As I described earlier, I sent him to slaughter and donated the meat to a homeless shelter in Glens Falls. I sent Harold, my other steer, and Luna, the beef cow, to a farm in Vermont.

I put a message up on a goat forum and a

young couple in Vermont came to take my goats to live on their farm. I gave twenty-five of my sheep to a farmer in nearby Warren County. I got rid of the six chicks my hens had given birth to. I sold my tractor, and a farmer came to take the round bales of hay away.

I sent the donkeys away along with the sheep. It broke my heart to do that — I was especially close to two of them, Lulu and Fanny — but I had lost faith in my own judgment and motives for loving animals. I could no longer differentiate between what I loved and needed, and the dysfunction and terror that had settled over me like a cloud. I realized I could never understand animals or love them in a healthy way if I did not sort out my own needs and emotions.

I got lucky. In a remote town where one might go to hide from the world, I found what I had really been looking for all along. I met a wonderful woman, an artist, and asked her to marry me. She accepted. Maria and I have lived together in love and connection. Lenore was a wonderful companion for me, but Maria is better.

There are things that animals can do, and things that animals cannot do, and understanding the difference is essential to knowing and respecting them. Once I found Maria, I began to understand an essential truth about animals.

Animals thrive in balance with people. They are not surrogate children, dependent and piteous creatures. Some animals — steers, for sure — are not pets. In our loneliness and disconnection, we increasingly view animals as a reflection of us; we bestow on them our ambitions, needs, yearnings, emotions. I had lost a true understanding of what they are like, what they need. Giant steers are not like puppies; they are something different. They deserve to be treated well, but they don't need cute names and treats. Goats need to eat weeds and brush, not cookies out of the hands of humans. I needed people and community in my life, not just animals spending much of their lives waiting for me to bring them goodies when in fact they didn't need the food and could forage and graze for themselves.

I wanted a human being in my bed at night, not just a sweet dog. I needed a community of people around me, human friends in bonded relationships. I needed to confront the reality of my new life, and my new life with animals — one that I could understand, afford, and manage.

Before my new wife, Maria, and I left our farm, we got Fanny and Lulu back. And then we got Simon, an abused donkey taken off a farm by the state police. He was the right animal for me. I loved him and helped nurse him back to health. He repaid me with gratitude and affection.

I was rebuilding my life and creating a new life with animals. Maria and I shared a love of animals and a passion for communicating with them. We taught one another how to know them and talk to them. We bought a home together. Our new farm was not a grand stage; it was simply the right place for us to live together. It was a small place, a sensible place, and we could afford it. It was close to a small town where we could make friends, begin to form a community. We were very happy to have our donkeys back, and after a year or so, they forgave me for abandoning them. We got seven or eight sheep. I got Red, a wonderful dog for us, calm, competent, and loving.

We could know these animals, share them with one another, learn to understand and care for them.

So the Great Collapse was a bridge for me. I was learning how to be healthy, how to be authentic, and in so doing, I was learning how to be healthy and authentic with the animals. I no longer had animals I didn't know, that I could send away, get rid of, exploit for my own emotional needs. These animals shared our lives, and our lives began to make room for animals who would teach us a great deal.

When we moved into our new farm, a blind old pony named Rocky came with it. We agreed to keep him, and Maria and I began to connect with the world of horses, an evolution that was soon to change both of our lives as well as my writing and work. Another gift of the Great Collapse.

A life with animals is a life of love and loss. They do not live as long as we do, and their lives are less secure. I choose to celebrate my life with these amazing creatures. I do not spend much time in mourning their loss; in a very important sense, they are not gone. They do not leave me, because they are always in my mind and my heart. Their spirits are always present.

I am lucky to know each one, to learn how to talk to them, love them, and be loved by them. To understand the boundaries between them and us, the dignity and respect they are entitled to. My journey with animals, my quest to understand them and communicate with them, came to life on the other side of the bridge.

7
Flo and Minnie: The Barn Cats

It is early morning. The sun is just coming up over the pasture, and it is brutally cold, minus-21 degrees. I am up early, going into my study to work. Flo, a cat who, like so many others, has come in from the cold, is sitting by the woodstove. She leans so closely into the heat that her brown coat sometimes turns orange in spots. We call her Crispy Cat.

I get a cup of tea, open my study door. Flo leaves the stove and comes into the doorway. Not here, I tell her silently, not in my office, I can't have a cat slinking around in here.

Okay, she tells me, I can try. Suit yourself. She flicks her tail, goes back to the living room, jumps up onto her favorite chair by the fire, and goes to sleep. I will not see her again for hours.

Cats are a new chapter in my life with animals. Flo was either a barn cat who made her way to us or a cat abandoned and tossed out of a car nearby. She hid for a long time. Now, thanks to our ability to talk to one another, we are very close. Flo has opened my eyes to the love and connection of cats. She has won me over in late middle age.

Flo is the first cat I have ever loved, and the first cat I have ever communicated with. She is one of the few animals I believe communicate with me instinctively and continuously.

Like many dog lovers, I never considered myself a "cat person," and yet I don't think I have ever communicated with an animal more easily, constantly, or clearly than I do with Flo. She is a mysterious cat who lived for some time secretly in the upper corners of our woodshed, who seduced me, got me and my wife, Maria, to feed her, taught me how to understand cats and speak with them, and now lives in our house in cold and wet weather.

As is often the case with cats, we will never really get the full story of Flo's life. Cats are the orphans and free spirits among domestic animals. They come and they go; some do not seem to really know how to live in our world — they run in traffic, get eaten by predators. Others seem to have a genius for survival. Flo is one of those.

When we moved into our small farm in

2010, we brought two barn cats — Mother and Minnie — from our previous farm. Mother was a fiercely independent and predatory cat. She lived in the barns, and I was mesmerized by her ferocity and her mysticism. She rampaged through the meadow across from the farmhouse, coming out with birds, rabbits, rats, chipmunks, mice, and moles, many of whom she joyously tortured before killing and eating them.

Minnie was a feral kitten I adopted from a waitress in a local restaurant. She lived in the barn with the chickens and slept near them. I think she thought she was a chicken. She was shy and timid, unlike Mother.

Mother kept an eye on me, but always from a distance. She would walk through the woods alongside me and the dogs, but she was always ready to vanish or bolt up a tree at the first sign of trouble.

It took several days to round up Mother and Minnie once we got them to the new farm. We put them in large crates in the big barn and fed them there for days so that they could become acclimated to their new home. Minnie adjusted quickly. She loved to sleep on top of the hay bales and make warm nests for herself, as she had done in her previous home. Mother, viscerally independent and with the wild, proud spirit of the barn cat, vanished almost immediately after we let her out of her crate and moved on, we think, to another home. We can't say for sure.

Soon after we moved in, we became aware that there was another cat around. We saw glimpses of a small brown tortoiseshell cat rushing under the front porch or up in the woodshed. She never came close to us, would never show herself.

We thought she might be living under the front porch; her tracks seemed to move in that direction. Or perhaps up high in the big woodshed at the back of the house. We climbed up and found a kind of nest up there, made of shredded paper and old rags. Judging from the indentations, a small animal was sleeping up there. But we never saw it.

We were never certain if this cat lived on the grounds, or stopped by, or had moved on, as Mother had. Months would go by during which we saw no trace of this visitor, then we might catch a glimpse of her darting across the yard.

She did not seem to want anything to do with us, and was careful not to be seen. I had this sense that she was watching, though, checking me out. Cats have a very interesting trait: they manage to be wild and domestic at the same time, both independent and needy. Unlike dogs, they never seem to completely give themselves away to people. There is always a part of them that says, "Not so fast, not so close. I get to decide."

There wasn't much we could do; nothing we really wanted to do. It was enough to care for the dogs, donkeys, sheep, and Minnie and the hens.

During one raging snowstorm, I was in the woodshed, stacking firewood into the back compartment of my SUV to transport to the house. I looked up and saw a small brown head staring down at me.

As I stared at the cat with big green eyes, I expected her to bolt or hide, but she stared back at me with a kind of impertinent look, as if to say, "What are you doing in my shed?" But also, something softer, something like "What are you like? Should I deign to get to know you better?"

I said hello, and then I started stacking the wood again. I did not grasp at first that I had just experienced the beginning of a kind of communication, of a brilliant and sophisticated campaign that was to upend my own fixed ideas about animals and then lead to a fluid and meaningful communication with an animal whose species was a complete mystery to me.

From that day on, Flo's head would pop up every time I came into the woodshed. I greeted her and she would stare at me and then vanish. I would chat with her a bit, ask how she was, talk to her while I stacked wood in the dark and cold shed. I thought she was talking to me. I got some images of the box where she slept, some feelings of cold, some sensations of being hungry.

After a few weeks, Flo started climbing down from her perch in the woodshed. She would sometimes come down a few feet, sometimes come level with me. One day I thought I sensed her calling to me. I leaned over and scratched her on the top of her head. She loved it; she purred and tilted her head and flirted with me.

We had begun our conversation. She was getting to know me; she was beginning to seduce me. We are going to have a relationship, she said. I intend to be your cat. I'm not really asking you, I'm telling you. I'm not a cat person, I said, but you are welcome to the shed.

I didn't see Flo for a few days and got worried about her. Maria and I searched the shed and the front porch, but there was no sign of her.

This bothered me more than I might have expected. I have strong feelings about barn cats being left alone to live their ethereal and somewhat mystical lives — many a night I've come into a barn and seen the cats dancing up in the rafters in the moonlight — and I have never been tempted to bring them into the house. Mother and Minnie made it very clear they had no interest in coming inside; they cherished their freedom. And my barn cats had always been lethal mice and rat killers; they kept the barn clean and rodent-free. I had no thought of bringing Flo into the house, but when she reappeared after a few days, I was relieved.

One especially bitter winter night, I brought some warm fish stew to the woodshed. Flo came sauntering down, sniffed the stew, ate it hungrily. I could hear the sound of her purring yards away.

Now I was talking to her. I spoke to her verbally, touched her and rubbed her, and now, with the stew, I had brought her sustenance and warmth. We were deep into it.

"You're not coming into the house," I said, again and again.

"Who would want to?" she seemed to reply, almost disdainfully, before vanishing up into the dark recesses of the woodshed roof.

Flo was intrigued by me. She was coming on to me, yet at the same time she made it clear that she was calling the shots, not me. She could take me or leave me.

But the thing is, she didn't leave me. She kept appearing, purring, presenting herself, staring at me with those green eyes. And I didn't leave her.

As the cold and stormy winter went on, I began bringing food more regularly. Most often she was waiting for me when I came to the shed. She seemed to know when I would appear. Every now and then there were gifts — a bird, mice, a mole or two — waiting for me on the woodpile. Sometimes Flo came down for cuddling; sometimes she just sat up high on the pile and watched me; once in a while, she didn't appear at all. I had no clue as to where she was or what she did when I wasn't there.

Flo and I settled into this routine, and after a few months, we were understanding each other. We were in sync in some strange and new way. Maria saw it right away. "You two just love each other," she said. "You are talking to each other all the time."

And we were. With my body, my emotions, and food, I spoke to her. And soon enough, with visualizations. The woodshed was a perfect place for visualizing — dark and quiet. I got a stream of messages from Flo. I saw a house with children. There were cows in the backyard, a barn. I had the sense of fear, an image of the porch. Suddenly, I knew where Flo came from. She'd been dumped by the front of the house, at night, I think. She became a barn cat and a porch cat. I saw an image of my dogs, barking, of a dark, safe place — the woodshed. And images of me, again and again, a warm glow around me, smells, feelings.

When spring came, I was outside more. Mowing, sitting in the Adirondack chairs, herding the sheep with my border collie, Red. Flo appeared next to the chairs. For the first time, she ended up in my lap. I was startled. I had never had a cat in my lap before.

Flo went from being a shadow, a phantom, to being almost ubiquitous. Whenever I was outside she would come out of the garden, out of the barn, down from the shed. I might be sitting in the chair, or on the rocks by the garden, or having lunch with Maria on the porch. Flo was always there.

If I was eating, she would lie down next to me. Sometimes I got irritated with her getting too close when I was eating, so she stopped. When I was done, she would climb into my lap, purr loudly, and go to sleep. I liked it. It was sweet and comforting having this soft and living thing in my lap. And then she would be gone, and we would go our separate ways. I liked the dynamic between us. Flo was not too clingy or too close; she seemed to sense when I wanted her around and when I didn't. She never made a pest of herself and had this extraordinary radar that seemed to alert her when I was ready for her. I would smile at her, close my eyes, and picture her asleep in my lap, and she would appear there.

I began to love this cat. It seemed we were perfect for one another. She was independent, and found a quiet corner of the living room — the back of a chair. She never bothered me when I worked, or asked for anything other than to be fed. Once in a while, she crawled into my lap when I was reading and purred softly. It was restful, meditative. I began to appreciate the skill and sensitivity of her campaign to win me over.

Behaviorists have long argued that cats, almost alone in the animal world, choose humans as their companions. They use their extraordinary instincts to sense the people who would best care for them. They are only partly domesticated, unlike dogs and carriage horses, but they are astute survivors who seem to grasp that humans are a source of food and safety. Unlike many dogs, cats are often very content being alone in a way that dogs are sometimes not. Cats spend the great bulk of their lives sleeping, they don't need the same amount of attention that dogs often require.

Flo had clearly chosen me. She tracked my movements around the farm and followed me, often at a discreet distance. When I sat and read in an Adirondack chair, I would look up and see her in the garden, a few yards away, basking in the sun, keeping an eye on me.

This became our new routine. If I sat in the chair, she came over and looked at me. If I shook my head and looked away, and imagined her staying away from me, then she did.

Flo never came to me when I wished to be left alone, but once in a while she would come close and stare at me. She was telling me she needed to be in my lap, needed some connection with me. And so I would oblige her.

It is important to understand the place of the barn cat in rural life. There is much controversy surrounding these wild creatures. Many animal rights groups believe it is cruel for barn cats to be left outside, never inoculated or given complete shelter.

Yet, if you have ever been on a farm, the issue seems much more complex. Wild cats are drawn to barns with livestock because there is often warmth, water, and grain there. Many farmers leave giant bowls of milk for their barn cats, yet very few farmers can afford to take twenty or thirty barn cats to the vet for shots every year, even if they could catch them. Money is never spent on barn cats. They get shelter and milk, but are otherwise on their own, living and hunting freely.

I embraced these conventions, with a few exceptions. I had only two barn cats, and I did catch them and take them to the vet for their shots every year. I would also leave food out for them on cold or snowy nights.

One night, Maria and I were awakened by the sound of a cat shrieking. It was a more urgent sound than we had heard before, and we both rushed outside, but could see nothing — no cats or other animals, no sign of a struggle.

We did not see Minnie the following morning and became concerned. She was always out trawling for food first thing after the sun came up. But barn cats often go off on hunting sprees, disappearing for days and then showing up when they need a warm or dry place to sleep. Later in the day, when we went out to the barn to do the afternoon feeding and chores, I heard Maria call out to me in alarm. "I found Minnie, I think she's dead!" I came running. Maria had seen one of Minnie's legs protruding from the sliding barn door. We rushed inside and found her. She was not dead, but close.

It looked like she had been attacked by an animal — a fisher or raccoon, perhaps — and had made it into the barn to hide. Her right rear leg was hanging loosely, almost as if it were disconnected from her body.

We rushed Minnie to the vet just down the road. We left her there for X-rays, but the vet said it looked bad. Something had grabbed and wrenched the leg so badly that there was massive nerve and muscle damage.

An hour later, the vet called us back. The leg was severely damaged and not treatable, she said. The options were to take her to a specialist in Albany — where they sometimes did elaborate and expensive reconstructive surgeries in cases like this — amputate the leg, or euthanize her.

Reconstructive surgery would cost many thousands of dollars, a minimum of $5,000.

Amputation would cost around $2,000. Euthanasia would be around $100. If we amputated, the vet said, the prognosis was good. Three-legged cats can't do all the things four-legged cats can do, but they can get around, adapt, and lead healthy, relatively normal lives.

I know there are some people who judge their love of animals by how much money they are willing to spend to heal them and prolong their lives and by how quickly they accede to elaborate and expensive treatments and surgeries.

I am not one of those people. While technically I am not a farmer — I see myself more as a writer with a farm — I have many farmer friends and feel very connected to them and their values. I respect these hardworking and honest people. I agree with their opinion that there must be limits on the impulse to save and rescue animals, if our understanding of their lives is to be genuine and our relationships with them healthy. Money does matter, and so does perspective.

I was unsure about what to do next, ambivalent about the choice we had to make for Minnie, who remained in a crate on sedatives and painkillers while Maria and I made up our minds.

Maria and I both instantly ruled out the surgical reconstruction. We didn't have five thousand dollars for specialists and complex surgery, especially when it was far from clear the leg would be saved. I could tell Maria was leaning heavily toward the amputation. She loved Minnie and because she knew I was not especially drawn to the cat, Maria was especially protective of her.

Aside from cost, I am rarely comfortable subjecting animals to prolonged and often painful medical procedures. The vet had assured us that cats do well on three legs and I had no doubt that that was true, but I was also certain that amputation would be a painful and disorienting process, especially for a barn cat. Minnie would never be able to run as fast or seek shelter as quickly as she had done before — that ability may have saved her life.

Maria and I went back and forth. I decided to try to talk with Minnie, to see if I could get a sense of her, a sense of what she wanted. I had often brought food to Minnie and she was comfortable around me, even if we were not close. I had rarely tried to talk with her, and yet here I was, faced with one of the most important reasons for learning how to communicate with animals. We don't have to always play God, to try to figure out what is best for a particular animal. Sometimes they can actually tell us, if we learn how to listen. They can help us reach good decisions.

The vet said she wouldn't be able to operate for a couple of days, so we had some time to decide. Just after the clinic closed, I showed up — with the vet's permission — and asked if I could go see Minnie alone. It was quiet in the holding room in the back. Minnie was sedated. She must have been in considerable pain. She lay facing me in a large crate with some food and a bowl of water next to her. Her leg lay limp by her side. She looked miserable but calm. Much of her food was uneaten.

I slipped a chunk of tuna fish through the slats in the crate. Minnie sniffed it and ate it quickly. The office was quiet, I got close to the crate and let Minnie sniff me. I wanted her to know what I was feeling. Mostly, it was confusion.

I may not have been particularly close to Minnie, but I did have an attachment to her. We had been through a lot together. She was one of the first animals I brought to Bedlam Farm, and I loved looking out the farmhouse window and seeing her perched up on a fence post, watching the meadow eagerly for any of the host of mice, moles, and rats who lived there. The life of a barn cat was rough, but in many ways it seemed wonderful to me. How many more-or-less domesticated animals got to live their lives in so free and independent a way?

Do you know that you've been lucky? I tried to ask her. Then I ran though some images in my mind, the operating table, Minnie with three legs. I didn't know how to project pain.

Minnie began meowing softly, and was focusing on me. She was smelling me, for sure, and she was also looking at me. I had the feeling she was gathering information.

We sat like that for four or five minutes. I thought I was probably kidding myself. The cat was sedated and in pain, and I was also ambivalent about what we ought to do. But then, as I was preparing to leave, I got a message. Minnie lifted her head, and she showed me the spark in her eyes. *There is more life in me. I'm not ready to leave.*

Minnie didn't speak to me in words, but she showed me images. I saw her sitting in the sun by the barn with three legs. I saw her hopping around the pasture. I saw her with mice in her mouth and some moles. I have sat down with a number of dogs and sheep — and one donkey, too — whose spirit was gone, who were exhausted by life,

210

who were ready to leave. Minnie wasn't. I went home and told Maria what I had experienced, what Minnie had said. We agreed to go forward with the amputation.

It's been more than two years since the operation was performed and I have to confess that I am as ambivalent about it today as I was then. It was a very painful procedure for Minnie and she has never regained the confidence or mobility necessary to live the life of a barn cat. I've watched her struggle to get to the high spots that cats love. I see her fearfulness of being out in the open too long, even on sunny days. She is essentially a house cat now.

During the day, she sits on the carpet by the woodstove or goes down into the basement — it looks like a barn down there; it is dark and quiet and safe and there are mice rushing around.

At night, she loves to curl up with Maria and sleep next to her. Minnie's wound has healed, and her body has compensated for the loss of a limb, but she can no longer live the life she was meant to live, and she suffered more than I am comfortable seeing an animal in my care suffer.

Maria has a different view. She believes the surgery was a success and Minnie has a full and loving life. While I may have a different perspective on Minnie's story, I do believe that she was not prepared to die, that her spirit was alive and she wanted to go on. She told me so. For that reason, I am glad that she is here with us now. Her message made a difficult decision much easier.

Flo's second winter on the farm was even more brutal than the first. Wave after wave of frigid arctic air sent temperatures plummeting to well below zero. Storm after storm dumped more and more snow around the farmhouse. We let Flo inside the farmhouse in January and she didn't leave until March.

The very first thing that Flo — a small and slight cat — did when she came into the house was to communicate with the dogs. We had three at the time: Frieda, a Rottweiler-shepherd mix, a hunter, a guard dog, a chaser, even a killer of cats (we were very worried about Flo being inside the house with Frieda); Lenore, a genial Labrador retriever, an old pal of Minnie since they had grown up together; and Red, a workaholic border collie and therapy dog.

Flo wasted no time in establishing herself. As we stood by nervously, Frieda came marching over to her, ruff up and tail and ears stiff. I moved closer. Frieda had caught and killed many a rabbit, chipmunk, raccoon, and skunk in her day — a few barn cats, too, I recall. She had been abandoned in the Adirondacks for years, and lived in the wild there before she was captured and adopted by Maria. If Flo couldn't handle Frieda, we would have to find a new home for her, and initially we were not optimistic. Frieda was brimming with prey drive.

As Frieda loomed over Flo, the cat turned calmly, hissed at the dog, and swatted her on the nose with her paw. Frieda froze, stunned, and then backed away. She never challenged Flo again, and for several months she would not even walk past her. She would go to any length — walking around sofas and the perimeter of the living room — to avoid her.

A day later, Lenore came bounding over to Flo, her big tail going back and forth. She too received a swat and a hiss, and never bounded at Flo again, although she would come closer than Frieda.

As for Red, a hiss was enough for him. Border collies don't need to be told twice. He never went anywhere near Flo. Over time, things became more peaceable. Red and Lenore would lie down close to Flo when she was sitting by the fire, but they did not bother her or approach too closely.

It was astonishing to me to see this little cat subdue, even terrorize, three large working dogs effortlessly. She quickly became the queen of the farmhouse. She slept where she wished, hogged the space in front of the woodstove, cleared a path when she walked to the kitchen or the litter box.

Here was another example of effective communication. The hiss was a clear statement to keep away, the claws a reminder and enforcement, her posture and demeanor a signal of confidence and dominance.

And that is how animals work. They don't traffic in human narratives. A human being would have assumed that Flo was too small to handle any one of these dogs, that they could have stomped her, run over her, or torn her to bits in a flash. Animals rely on a different kind of logic, based on a different system of communication. Smell, posture, images, demeanor, eye contact — these speak volumes to animals, and we can use the same methods in our own interactions with them.

After she got the dogs firmly under control, Flo took over my reading chair, a big stuffed chair next to the woodstove. I didn't want her sitting in my lap or bothering me in the other chairs in the living room —

where I ate or talked on the phone or looked at the news on my iPad. She came up to me once when I was reading there — I got uncomfortable, bristled a bit, thinking she would be pestering me all through this long and hard winter.

I am certain she sensed, or perhaps smelled, my irritation. My body language shifted. I crossed my legs, turned away from her, put a book in my lap.

From that moment, she has never approached me in any chair but the big stuffed one, and this is our deal: when I am not in the chair, she can sleep in it; when I am in the chair, she can lie in my lap. She didn't sit on any other piece of furniture, and won't bother me when I am anywhere else in the house.

Like many people with complex childhoods, I was put off by cats, by what I thought of as their "slitheriness." It made me uncomfortable to have a cat rub against my legs or jump up on me unexpectedly.

Flo never does any of these things. Our understanding is clear; she senses my comfort level and respects it, as I respect her need for human contact, her desire for a warm lap to curl up in.

We were both changed by this relationship. Flo and I met in a place where we were both comfortable. And we talk all of the time. I have used all of my tools to communicate with Flo. I've also studied up, poring through scores of journals and behavioral studies on animals, looking for information on the particular ways that cats communicate. Not surprisingly, they are different than dogs.

Dogs are pack animals; they travel in groups. Cats are solitary; they move and hunt alone. Dogs are omnivores; they can eat plants and meat. Cats are carnivores; they survive on meat and prey.

Because they often live and hunt alone, cats make more decisions than dogs, so their minds and instincts often develop more intensely. Because they eat more often than dogs, they also need to make decisions more frequently.

A cat is generally more comfortable being alone than a dog. Some dogs are more prone to being anxious when they are by themselves, if they are not used to it. It is very difficult to measure the intelligence of an animal, but behaviorists do say that cats communicate much more intuitively and indirectly than dogs. Dogs seem easier with people and respond to direct commands. But cats intuit — they sense who is their friend and who is not. They are less trusting. Dogs approach one another — and people — more directly. They sniff, even lick one another, communicating through eye contact, tail posture, and body language.

Like dogs and donkeys, cats smell emotions in humans, and look to body language for cues. Flo will watch me to see where I sit before she decides where to go. When she wants to go outside, she doesn't bark or wag her tail; she goes near the door and circles, sending a psychic signal — I believe she is. I can be sitting around the corner in the living room and I can sense when she is at the door, wanting to go out and hunt.

I do not assume Flo knows every word I am saying, or even many of them. But I believe we talk to one another. We work out the details of our relationship, telling each other what we need and don't need, what makes us comfortable or uncomfortable.

I see Flo very much as a partner, and I have grown to love her in the same way I love my dogs. I now understand the independence that cats feel and project, and why that is so appealing to so many people. I grasp their intelligence and intuitiveness, which often seems superior to that of dogs.

There is a sense of entitlement about cats that many dogs do not have. Flo takes her role for granted. She does not work for it, demand it, or wag her tail or chase a ball for it. She takes it as her birthright.

I believe there is a spiritual dimension to talking to animals like Flo. If you see them as piteous or dependent or try to talk to them in words they don't understand, if you approach them in an atmosphere of anger or frustration, then they almost visibly close up, and turn away.

Animals are in many ways alien creatures, but there is a point, a contact point where we can break through to them, and they to us, if we are willing to open our eyes, ears, and minds to the wiser understanding, to the more mystical nature of that contact.

Cats are very different from dogs.

They are generally more independent and less needy, less demonstrative in their affections, and harder to communicate with directly. They are not pack animals, so they are not as social and are much more difficult to train. People love them for their aloofness, an interesting counterpoint to dogs, whom many people love for their ability and desire to be close to us.

With cats, there is the sense they are not giving it away, that you have to work for it. But the intense devotion and attachment people have to their pets is the same whether it's a canine or feline.

8
SAVING THE ANIMALS

Ariel Fintzi is a legend in Central Park. The homeless men who sleep in the park call him an angel because he gives them food and money. Some of the lost boys of New York, most of them African-American or Latino, come to Central Park from the outer boroughs to talk to Ariel. They pet his horse, Rebecca, solicit his advice about staying in school or off drugs, ask him for a loan.

He is Mr. Fintzi to them and some will tell you that he has saved their lives.

But it's not just the homeless who rely on Ariel. Every morning, Jane, a ballet dancer left a quadriplegic by a nerve disorder, calls Ariel on his cell phone and asks to meet him at the Bethesda Fountain in the center of the park. Ariel, she says, will not accept payment for his time and is always — always — waiting for her when she arrives. She looks up in her wheelchair into the eyes of Rebecca, Ariel's horse, and she says these visits are what she lives for.

Ariel is a New York City carriage driver. Born in Israel, he grew up with horses on a kibbutz. He is a profoundly spiritual man and believes horses have great power to heal and help human beings. Ariel now lives in the Bronx, but his real home is the park. He is there as often as he legally can be. He knows every path, pond, tree, and rock formation. If you are in love, or touched by the stars, Ariel is very happy to take you on an enchanting midnight ride in Central Park. The ride will remind you that the horses carry magic and mystery with them, and if they are taken from the park, it will lose its soul and spirit.

Ariel is a friend now, and people who consider themselves animal rights activists have accused him and other carriage horse drivers of being hateful and greedy, torturing and exploiting their horses by forcing them to pull carriages in the great park. This confounds Ariel, as it does people who know him.

If you are curious, you can go to the park the next time you might be in New York City — any driver can tell you where Ariel is — and watch him with Rebecca, his carriage horse. He sings to her in several languages, plies her with carrots and cookies, comes to the stable before dawn to brush her and talk to her and tell her stories. Rebecca is a proud, loving, healthy, and eager-to-work horse. Ariel trained her himself. You could blow up a car right next to her, he says, and she would not kick or bolt.

I never expected to be drawn into Ariel and Rebecca's world, to be drawn into the carriage horse controversy, but it proved to be one of the most profound experiences of my life with animals.

In January 2014, I read a series of news articles quoting various animal rights organizations and activists as saying that the more than two hundred draft horses of the New York City carriage trade were being abused and that the trade ought to be banned, the horses sent off to rescue farms.

Like most people, I was only peripherally aware of the carriage horses. My only connection to the horses was seeing them lined up on Central Park South, strikingly beautiful creatures standing calmly in a rolling sea of tourists, pedestrians, dog walkers, office workers, bicyclists, joggers, pedicabs, and picnickers. I had never been on a carriage ride or thought to take one. That was something tourists did, like going up to the top of the Empire State Building. I would stop once in a while to look at the horses or pet one on the nose, as so many people do. I thought they were quite beautiful and fit very naturally into the park.

Protestors had become increasingly vocal and angry about the horses. The stories painted a grim, even horrific picture. The media reported on their claims, almost invariably without confirmation or reply from the carriage trade. It was alleged that the horses were confined in stalls too small for them to lie down in, that they were fed rat-infested grain, stood in filthy stables in mounds of manure.

The horses, said the demonstrators, were depressed, lame, covered in sores, and suffering from lack of proper medical care. They were overworked, sickly, treated harshly by their handlers, who were described online as marginal people, alcoholics, thieves, greedy, money-grubbing villains. The drivers and owners of the horses were consistently portrayed as evil, subhuman. They were not businesspeople, or misguided people, or even old-fashioned people, but something other than human beings, not entitled to any measure of respect or consideration.

The local animal rights groups seeking to end the carriage trade had recently picked up some heavy-duty allies in their campaign against the carriage horses. The ASPCA and Humane Society, once guardians and caretakers of the carriage horses in New York City, had joined ranks with the new and more extremist animal rights groups — PETA, the U.S. Humane Society, NYCLASS — and had changed course. They said the horses didn't belong on New York's congested streets any longer, that they ought to be returned to the wild or to rescue farms, where they would never have to work again. The campaign against the horses really heated up in December 2013 when the new mayor of New York City, Bill de Blasio, said it was "inhumane" for the horses to be in New York City. "We are going to get rid of the horse carriages. Period," de Blasio said in response to a question at his first press conference even before taking office. He said the horses would be banned in the first week of his term, on "day one."

"Just watch me do it," he said. "It's over."

NYCLASS, the group funding and spearheading the ban effort, had given the mayor enormous financial support in his election campaign and devoted nearly one million dollars to defeating his main opponent in the race.

New York is a cynical town, and many

people in the carriage trade and their sup-
porters came to believe that the mayor was
supporting the real estate developers —
including the head of NYCLASS and leader
of the ban-the-horses-campaign — who
were said to be trying to acquire the horse
stables in the hot West Side real estate market.

Why did I care? I write about animals, of
course, but something else has become clear
to me in the last couple of years. I think the
horses called me there. At least that's what
a Native American spiritual leader told me,
and what I have come to believe. The horses
imagined their own future.

My own idea was different than the under-
standably suspicious carriage horse drivers.
It has become a popular idea in America,
this notion that it is cruel for working
animals to work with people, exploitive for
animals to uplift or entertain people.

Given how little urban dwellers know
about horses or how rarely they see them, it
seems almost inevitable that they would
come after the carriage horses. There is no
concrete evidence that real estate development was the only reason for the campaign
against the horses. The campaign marked
the first time in modern American history
that a big-city mayor was a member of an

animal rights movement considered too extreme and militant for most animal lovers or for mainstream politics or politicians. The mayor, who admitted that he had no experience of any kind with animals, came to fully accept the logic and rhetoric of the animal rights movement in his city: it was cruel for horses to be in New York, work in New York, travel the congested streets of the city.

He seemed to find the people in the carriage trade repugnant, outside of the moral community of good citizens. A self-declared populist and progressive, he refused to speak with any of the drivers or owners. He and his aides would not meet with them, lunch with them, visit them, or consider their arguments. He refused to worry about their jobs or respond to the continuous harassment the drivers faced in the city.

De Blasio told one carriage driver who brought his young son to one of the mayor's public appearances and who approached the mayor to ask him why he was seeking to ban the carriage trade that he thought the driver's work was "immoral." Historically, the carriage drivers were among the first members of the Teamsters Union, and the union went to bat for them when they were very much alone and in peril. The mayor was determined. The drama of the New York

City carriage horses had just begun, for them and for me.

I called Christina Hansen, a spokesperson for the carriage trade, and asked if I could come and see the horses for myself. Christina, a passionate lover of carriage horses, had given up an academic teaching position and a job as a horse carriage driver to come to New York and fight for the horses.

She was quite open. She said I could go to the stables anytime I wanted and see anything I wanted. She didn't sound like someone who was hiding something.

I would talk to Christina often in the next couple of years, peppering her with questions. We sometimes argued about what I wrote, but she never once failed to respond to my questions, and in hundreds of interactions she never once told me a lie, distorted the truth, or covered up a single thing. I came to know her well and trust her. She was unwaveringly accurate and truthful.

I wanted to write about the horses. I was drawn to the idea that domesticated animals were vanishing from the world, many at the hands of animal rights groups that believed it was exploitative to make them work for people.

As someone who had lived with working animals — dogs, donkeys, horses, sheep —

for years and written books about them, I believed the carriage horse story was important, and had enormous implications for animal lovers and the remaining animals in the world. New York City is America's biggest stage in so many ways. If the horses could not stay there, then there would be very few places in urban and suburban America — where 90 percent of Americans now live — that they could remain. And then, there was this communication thing. I felt like I was being called to go there. I would soon understand that feeling in a deeper way.

Maria and I had inherited a pony when we moved into our new farm, but before that I had never had horses or written much about them. I didn't grasp until later that the carriage horse issue was a very spiritual one for me, one that cut to the core of my philosophy about animals. I realized that everyone seemed to be talking about the horses, but no one was talking *to* them. And, very reluctantly, I came to realize that they were talking to me

Hansen told me I could come to the stables anytime, day or night. I could come in and see any horse I wished, with or without advance notice.

Maria and I boarded a train early one

Saturday morning and traveled to New York, where we took a taxi to the Clinton Park Stables on West Fifty-Second Street and met Hansen and Stephen Malone, a veteran carriage driver and the son of Paddy Malone, one of the founders of the modern carriage trade.

I am no equine authority, but I've been around horses and other animals for years, on my farm, in my travels, in my photography work, in my rural Upstate New York county. My wife and I were both somewhat shocked by what we saw in the city. The horses I met on this trip were calm and well groomed; many were lying down, something horses only do when they are truly relaxed.

The stables were clean. The horses had room to turn around in their stalls. They had good hay and clean running water. There was heat in the winter, fans in the summer. The harnesses and tack were spotless; the animals were alert, friendly. The grain was carefully stored and well kept. There were no traces of mouse or rat feces.

The horses showed none of the signs of abuse or neglect that are all too familiar to me or to any horse lover or equine vet — skittishness, unease around people, ears laid back, the whites of the eyes showing, sores, rough and dry coats. This impression was reinforced by a score of behaviorists, equine veterinarians, horse lovers, and trainers who, like me, came to New York to see the horses for themselves and unanimously found them to be healthy and well cared for. There is now a vast archive of studies, reports, findings, and observations about the New York carriage horses. They all say the same thing: the horses are the luckiest horses in the world.

Looking around, we saw many drivers coming and going. They were especially affectionate to the horses, who seemed excited to go to work, eager to be harnessed, happy to come out of their stables, walk down the ramps, get hooked up to their carriages. These horses all exhibited the eagerness characteristic of working animals about to do their thing — I had seen it in other horses, in my border collies, in therapy dogs, Labs in the woods, donkeys guarding my sheep.

I was surprised, but still wary. I had been a journalist for many years — the *Washington Post, Philadelphia Inquirer, Boston Globe* — including as an investigative and political and police reporter. I was viscerally suspicious of information given me, and I knew how to filter through it. I do not take anyone's word for anything.

Maybe this was a show put on for people like me? If the mayor of New York was so adamant about banning the horses, there must be something to it. And I had not been immune to all the ugly accusations: horses killed in traffic, horses endangering people, sick horses working until they died, lame horses, horses with sores, starving and freezing horses, overheated horses. I was prepared to see some sorry and pathetic animals, not these gracious, healthy, and very obviously well-cared-for and trusting draft horses.

Anyone one who has been around animals for even a short period of time can easily spot abused animals: they are wary, disoriented, sluggish, often emaciated. Their coats and skin are a mess, their claws and hooves too long, their teeth untreated. They are highly suspicious of people and invariably either restless or lethargic. There is always some sign when an animal is abused. Visiting with these horses, I saw none.

Abuse is a concept that has been so exploited and misunderstood that it almost has no real meaning anymore. Legally, an animal is abused when it suffers grievous injury or death as a result of human neglect or cruelty. Abuse is not an opinion on Facebook or something to be suddenly and arbitrarily defined in press conferences or on blogs.

Whether the horses belong in the city or not, it is safe to say that they are not abused. They are treated as well and carefully as horses can be treated, far better than most horses in the world are treated.

There is a carriage horse in New York named Spartacus who is much loved and cosseted by his driver, Tony. If you ask the other drivers about Tony and Spartacus, they will tell you that Tony sees Spartacus as a member of his family. Tony talks to his horse, sings to him, brushes and combs him, showers him with treats and kisses. The other drivers joke about Tony, about how he calls Spartacus his "baby" and dotes on him like a proud papa.

One warm summer day in 2014, the carriage in which Spartacus was harnessed got tangled with another horse carriage and tipped over, causing Spartacus to fall to the ground, where he lay on his side. Hundreds of people were nearby; the scene was captured on smartphone cameras and shared all over the Internet. Several people who read my blog were present. As is often the case in accidents and other public dramas, many people were not sure how to understand what they were seeing. Many people had never seen an actual horse before, an animal familiar to them only through books, movies, and TV.

When a horse in a harness falls over, horse people know what to do: how to be calm, to talk to the horse and keep him or her still, so that they can untangle the harness and get the horse on its feet safely.

If a horse tangled in a harness gets to its feet too quickly, it can panic, break a leg, injure people nearby, lose its life. Grounded horses stay calm; high-strung horses can freak.

If the rider or driver knows how to talk to his or her horse, then the outcome is usually good. He will get down next to the horse, stroke its neck or mane, talk to the animal quietly and gently, tell it to stay and be steady, keep the animal still until the harnesses can be removed and the horse can rise safely to its feet.

Carriage horses are, by nature, among the most placid domesticated animals in the world. Although they are prey animals, they are famously calm under pressure, easy around other species, and very connected to their humans.

Fortunately for Spartacus, Tony talked to him all the time. As the big black horse lay on his side, tangled in his harness, Tony lay on the ground, talked with him, stroked his neck and mane, told him to be still.

It is upsetting for people to see a fallen horse, and the carriage drivers, long criticized by animal rights activists who believe the horses are being abused, knew there might be trouble. In New York City, hundreds of people are injured in motor vehicle accidents — many fatal and grievous — every single week, but such incidents rarely make the news. If a carriage horse stumbles or limps, it's major news for weeks. The carriage drivers are well aware that every movement they make will be scrutinized, captured on cell phones or video cameras, seized upon by their critics.

It took Tony and some other carriage drivers about two minutes to remove the harness and untangle it from the big horse. Then Tony told Spartacus to get up, which he did. He was taken back to the stables, pronounced fit by a police veterinarian, and then sent to a farm in New Jersey for a week so he could rest and be examined further. He was fine, and returned to work, where he works still in Central Park, with Tony.

A number of animal rights groups predictably seized on the incident to argue that Spartacus and the other horses ought to be banned, the carriage trade shut down. They said it was cruel to force a horse to lie on the ground for so long. They alleged that Spartacus was forced to go back to work rather than get checked by a veterinarian. They said Spartacus was spooked by a passing bus, and accused Tony of caring more about the harness than the horse. They assumed the care he and the other drivers took in removing the harness was an effort to save it at the expense of Spartacus. This accusation was widely reported and repeated for days in the media and online.

These reports were untrue — I talked to police, witnesses, transit authority officials, carriage drivers — and caused Tony to burst into tears at the thought he would ever harm his beloved horse. (Tony pointed out that Spartacus, who feeds Tony and his family, is worth a lot more than the harness.) He said the people accusing him of cruelty had stolen his dreams, the dreams of an immigrant who came to America to live safely and be free.

In a sane world, Tony would have been praised for his calm and quick response. He made sure Spartacus and the people around him did not panic and were not harmed in any way. As it was, he was reviled for days in the media and on the websites of several animal rights groups.

We are living in a critical time for animals. Animals' natural habitats are disappearing, demolished by human greed and development or by the growing ravages of climate change. Many animal rights organizations are demanding that the carriage horses and circus elephants and ponies in the farmer's markets be returned to nature and to the wild, but there is now no nature for most of these animals to be returned to.

Pope Francis speaks of the need for a new understanding of animals in his powerful encyclical, *Laudato Si'*, one of the most influential documents relating to the earth ever published:

> If we approach nature and the environment without this openness to awe and wonder, if we no longer speak the language of fraternity and beauty in our relationship with the world, our attitude will be that of masters, consumers, ruthless exploiters, unable to set limits on their immediate needs. By contrast, if we feel intimately united with all that exists, then sobriety and care will well up spontaneously.

Sobriety and care are two wonderful

concepts in support of our need for a new and more mystical understanding of animals.

It is not enough to say that animals like the carriage horses must be suffering if they work with people and earn money. Working with people and earning money are both essential for domesticated animals if they are to survive in our world.

In fact, the carriage trade in New York has achieved what might seem on the surface to be impossible. They have found a way to keep animals like horses safe and healthy and engaged in the most densely populated geography in our country. They have created a new habitat and environment for them even as we have taken no responsibility for destroying their original and natural homes.

Consciously or not, they have paved the way, shown us how animals can be saved and kept healthy and relevant in the modern world. The politicians and the people who claim to speak for the rights of animals do not yet seem to grasp the real significance of the horses, their real message to us.

All over the world, forward-thinking municipalities and young farmers are looking to domesticated animals — mules, donkeys, draft horses — to replace automobiles and tractors, to reduce carbon emissions, lower costs, and reduce the use of fossil fuel. Beyond helping our bleeding planet, working with animals is a much more rewarding experience than working with machines. Farmers love their cows, carriage drivers love their horses, people love to see and touch and feel animals wherever they are. It's hard to love a tractor or a pickup truck the same way you love a dog or horse (although I sure loved my farm truck and tractor).

They are vastly more costly and destructive to the fragile environment than horses or elephants or donkeys.

Saving animals is an urgent moral and environmental issue. Instead of turning horses away from New York City, we should be emulating the model of the carriage trade and looking for ways to keep them here, to relieve congestion, reduce pollution, advance the idea of community and connection. And to preserve the ancient bond between people and domesticated working animals. It is not progress for working animals to be denied work; it is the new abuse.

For several years, I attended numerous demonstrations in the city against the carriage horses. The demonstrators in New

York do not speak with animals or listen to them. They will not touch a horse or bring a carrot. They do not seem to know that the big carriage horses have never lived in the wild, and could never survive there even if it still existed. Draft horses are big animals — they need about four bales of good, fresh hay a day to survive; they need to have their hooves trimmed, to be treated for parasites; to be kept out of extreme weather; to walk on roads without big holes; to be kept away from snakes and predators and starvation, protected from fighting with one another over dominance and food.

The horses need to give human beings an incentive to keep them around and keep them safe and healthy. All of these things are present in the lives of the carriage horses. They live long and healthy lives doing safe and regulated work with people who love them and take responsibility for caring for them. They have moderate work to do and are beloved by countless people, who flock to ride with them, give them carrots and apples, touch them and photograph them.

What more do we want for animals than this in our changing world?

In a sense, the same reality applies to elephants in the circus. Few animals have suffered more in the wild than Asian elephants. They are relentlessly hunted by sportsmen and poachers; they are pushed out of their homes by flooding and climate change. The few preserves available for them are notoriously underfunded and poorly run. Because an unknown number of them have been mistreated or abused, almost all of the elephants have been driven from their work in circuses. They have nowhere to go, no one to care for them, no future in our world.

It is naïve to think that somewhere there is a pasture or preserve for all of the elephants to retire to (or the big horses), where they will be fed the enormous and staggeringly expensive amounts of food and nourishment they require to survive.

Elephant sanctuaries estimate that it takes between 300 and 600 pounds of food to feed an elephant each day. Is it really sober or caring, to use Pope Francis's benchmarks, to believe that the hundreds of elephants being driven out of circuses in the United States and Canada by the people who claim to speak for the rights of animals will all find homes where they will be fed that amount of food at private expense for the rest of their lives?

Those familiar with the long history of

Asian elephants and their work with people know these animals powerfully attach to human beings, that they need work for their own grounding and health, that many things that seem cruel and abusive to people are not cruel and abusive to working animals.

If we talk with them, we can know what is good for them. If we make remote, often politicized and emotional assumptions about them, if we insist on seeing them as versions of us, if we lose the sense of awe and wonder that contact with them inspires, then we are condemning them to extinction.

The carriage horse controversy was a turning point for me and my understanding of animals. I asked myself: What did the horses really need? What did safety mean for them? What is abuse? Is it really cruel for working animals to work?

I don't mean to suggest that all carriage drivers are perfect, saintly people. I know a lot of them by now, and their personalities run the gamut just as in any other profession. Some are nice, some reserved. All those I've met are animal lovers, though. They enjoy working with the horses, riding them, watching the tourists and children fuss over them. They carry bags of carrots, and buckets of oats. They trot repeatedly over to the horse drinking fountains that are such a lovely feature of Central Park. They talk to the horses all the time. There is rarely any trouble with these big and beautiful animals. Most of the horses follow their routes in the park with very little human direction.

The animal rights people say the horses are lonely, that they need to be socialized, but the truth is that the horses are never alone. There are always other horses near them. They communicate with each other all day and all night.

There is something knee-jerk about our current obsession with animal abuse and our insistence that animals be given the same rights as human beings. And who, in our time, even really knows what that means? I met an older cop in New York City who rides a police horse — he says the big difference in the perception of animals comes from the fact that people making decisions "in their best interest" no longer know anything about them.

It's a good point. The mayor of New York, who wishes to ban the horses, has never ridden one. Nor can he recall ever taking a carriage ride. The president of the City Council, who is eager to ban the horses, says she is qualified to speak about them because she has two rescue cats.

At the time, the media almost daily published claims that the horses were unsafe in New York, that they did not belong in the city, but when I talked to the police, they told me that no writer or reporter had called them in years about the horses. I questioned them about horse fatalities and was told that in the past thirty years (and millions of rides), three horses had been killed in traffic accidents.

The carriage trade is one of the oldest businesses in New York City. It was founded more than 150 years ago when Central Park was built. Frederick Law Olmsted, its designer, believed the carriage horses were so essential an element, he built many of the bridges and paths just for them, so that New Yorkers of all backgrounds and incomes could see the horses and mingle with them in the great park.

The trade has always had close connections to the immigrant population. People in Ireland, Italy, and other countries had lived and worked with horses for thousands of years and knew them well. It was natural for new immigrants to seek out work in one of the stables, starting out as stable hands and working up to medallion-holding carriage drivers.

It was a comfortable way for immigrants to come to the United States, find work that they loved — outdoors and with animals — and support their families. In the entire history of horses and carriages, no one had ever considered it cruel for workhorses to pull light carriages, especially on flat ground. Driving a horse carriage was one of the least controversial jobs available anywhere. Suddenly, in modern-day New York, it has become one of the most controversial. The drivers are embattled, the subject of demonstrations and protests, accusations of abuse and thievery and cruelty made almost daily against them, their way of life and livelihood under siege including from some of the most powerful people in the city.

There are three different stables in New York City, all on the suddenly valuable property of the West Side of Manhattan, in what was until recently a dingy and impoverished warehouse district. As the district has gentrified and drawn the attention of developers, there is suddenly enormous pressure for the stable owners to sell, and growing questions about where the carriage horses can be stabled in the future.

The carriage trade is perhaps the most intensely regulated business in New York City. Five different city agencies oversee the care of the horses, from the police department to the health department. Drivers are licensed, stables inspected regularly.

Every aspect of the horses' lives is regulated. There are hundreds of pages of regulations governing their welfare. They get five weeks of vacation a year, and can only work carefully prescribed hours — they can't start before a certain time, or work past a certain time (it varies depending on the day of the week). They are inspected regularly by city and police veterinarians and cannot work if they are lame or injured in any way. They can't work in extreme heat or cold.

When I became involved with the carriage horses, nothing in my life or work until that point had helped me to see so clearly how our culture has lost any connection with the animals left in our world, or with the natural world itself. I was to see clearly the profound failure of modern media to deal with complex issues like the future of animals in our world, and to verify or investigate vicious and hysterical claims made against innocent people. I became alarmed at the great dangers facing the animals because people no longer understand them or know what they want or need.

The carriage horses led me many places.

They were responsible for a rekindling of my life as a journalist, of my passion for facts and truth. They took my writing about animals in a new direction and set me on a quest to achieve a new and more mystical understanding of them. Through the carriage horses, I was to make some of the most powerful personal connections of my life. I came to know and love many of the carriage drivers, and their suffering and persecution was to break my heart and inspire my work.

These drivers and their horses connected me in the most elemental way to the very spiritual idea of talking with animals.

There were other troublesome things about the campaign to ban the carriage horses from New York City. The people who know horses best — horse lovers, trainers, vets, farmers, biologists — were never consulted or interviewed by reporters, nor is there any record of the mayor or anyone on the City Council talking to them either. It was almost as if expertise and science no longer mattered.

The animal rights protestors that I encountered — and I talked to many at several different demonstrations — did not seem to know that there is no longer any wild, and that big draft horses have always lived in cities and on farms, that they have never lived in the wild and would not last long there.

"You can see the horses are lonely," Carol, an office secretary, told me outside the Plaza hotel one morning. "They should be running free, out on pasture all day, with their friends."

They did not know that horses who stand with their heads down and leg cocked are not depressed or lame but relaxed and safe.

There is no evidence of respiratory disease in any of the carriage horses, despite activists' claim all are suffering from the city's traffic fumes. No human being in New York City has ever been killed by a carriage horse, not one in 150 years. Jared Diamond, one of the world's most famous biologists, has said that draft horses are the most domesticable animals on the planet: gentle, adaptable, affectionate, tolerant of other species, eager to work in crowded, noisy circumstances.

Buck Brannaman, the world's most famous horse trainer, inspiration for the bestselling novel and Robert Redford film *The Horse Whisperer,* says the carriage horses are the lucky ones, they have good work to do and good care. The abused horses are the ones languishing on farms with nothing to do but eat and drop manure — the fate the mayor has chosen for the carriage horses.

Beyond the eternal arguments that swirl about the care and abuse of animals is the very sad reality that we have become utterly disconnected from the animals in our world. Between runaway development and climate change, they are vanishing at a horrific rate.

The carriage horse controversy led me to a better way to treat animals. To understand the real lives of real animals. To communicate with them and listen to them. We don't have to fight with one another about what is best for animals; we can see them, live with them, know them and talk to them.

The New York City carriage drivers have found such an understanding. So have many elephant trainers, yet they have been reviled and censured for it.

We are banning the very animals we can and should keep in our midst because we have lost any sense of fraternity with them. We can no longer see them in context of their reality, only through our own.

Spartacus's fall was a turning point for me, pointing up the urgency of talking to animals and respecting the humans who do. We can make better decisions about animals if we know them, talk to them, listen to them. If we see their need for us, and our need for them, if we can know, to some extent, what they feel and want, then we have finally come closer to that new understanding.

If we only see them as fragile and dependent creatures who must be saved from human beings, then our attitude will continue to be that of masters, consumers, ruthless exploiters, and emotional parasites. We will be unable to set limits on their real and immediate needs, and they will continue to vanish from our lives.

What does Spartacus really need? To be released into nature, confined to stand and drop manure all day on a rescue farm? Or does he need to partner up with Tony, bring sustenance and health to both of them, to engage in meaningful work that brings joy and happiness to humans?

Part of truly listening to animals means understanding that they're not all the same.

There is a carriage driver — he asked me not to use his name for fear of recrimination — who went to Pennsylvania to save a big draft female horse from being sold to a butcher at auction. He rescued her and brought her to New York. She was his fourth draft horse. Each of his previous horses had loved her work in the park, loved her life in the stables, loved all the attention from children and tourists.

But the new horse was different, he said. She was shy around the other horses. She wasn't aggressive but didn't seem to like being handled. She made her way calmly through the busy streets of New York, but he didn't sense the same comfort and ease he had felt in his other horses. Standing in line waiting for riders, she was skittish, he thought. She didn't like the buses that went by, or the constant honking of the cabs.

"She was a farm animal, I think," he told me. "Most of the horses love the life in the stables and in the park, they love being around the other horses, they love all the attention and they love to work. This one was different, she just didn't seem at ease to me."

So, he sold her to a gentleman farmer in northwestern New Jersey who uses her to haul firewood and help plow his hobby farm. As wary as she was in the city, she took to her new life. She was a one-person horse, the driver thought, and she just didn't like the life in the city.

I learned from all the dogs, horses, sheep, donkeys, chickens, and cats I have lived with that it is wrong to generalize about animals.

They are not all the same. Just as every human being has a distinct personality, every animal has its own genes, instincts, behaviors, and environmental influences.

Fraternity means a kind of partnership, not emotional or physical domination. Thousands of years ago, we saw animals as beasts of burden who existed only to serve us; we owed them no mercy or compassion. Organized religion began to change the world's view of animals. Jesus Christ was perhaps the first animal rescuer in recorded history. He saved a puny donkey a farmer outside of Jerusalem wanted to kill and rode it into the fabled city. St. Thomas Aquinas argued that human beings must be merciful to animals so they can learn how to be merciful to people. Muhammad urged his followers to treat animals gently, and the ancient Hebrews decreed that animals must be given a day of rest on the Sabbath, just like people.

The pendulum has swung far and wide, back and forth. In our time, it sometimes seems that people worship animals and love them more than they do their fellow human beings. There is, as of this writing, no national rescue movement involving the re-homing and care of needy people.

As the idea of animal rescue grows by the

day, the idea of rescuing people seems less and less popular. Perhaps the two are not unrelated. It is not always easy to love people; it is simple and sometimes more rewarding to love animals.

Pope Francis and his encyclical may popularize a new understanding of our relationship with animals. His ideas reflect the vision of Henry Beston, who wrote that animals are our partners in the world, not our siblings or dependents. They share the joys and travails of the world. We do not owe animals a perfect life, or a better and more protected life than people enjoy. We cannot guarantee them a life free of challenge and surprise and pain any more than we can promise it to ourselves.

We are at a crossroads in our world — if the animals are to be saved, we need to get busy doing it, and if people are to be saved, we need to actively engage in that effort, too. That is fraternity writ large.

Saving the animals means reawakening to the idea of them as precious and individual creatures. If we do not know them, we cannot save them. If we do not learn to understand them and communicate with them, we cannot learn how to treat them and discover what it is they really want and need.

The animals of the world have been betrayed by the very people who claim to speak for their rights. Soon the horses and the elephants and the ponies will be gone from the world, and our children and grandchildren will ask why they were driven off and where they have gone.

We cannot keep killing animals, hiding them away, putting them in danger in the name of saving them and loving them. It is urgent that the people making decisions about the future of animals know something about them, and can communicate with them. The stakes could not be any higher.

Lucky, Stanley and Julius, Orson, Elvis, Flo, and the carriage horses and the donkeys and sheep have all led me, step by step, to this new understanding, this new way of looking at animals and talking to them and listening to them. It has been a journey of a thousand steps. I remember Lucky telling me for the first time that I was safe, and could be loved, and could love, and could be strong. I remember him telling me how little he needed or wanted, just a kid to attach to, a need to fill, a wound to heal.

I remember listening to Orson, and hearing his plea for relief, relayed to me and repeated by a shamanic healer with a gift for talking to animals.

And I remember Elvis telling me how

much his legs hurt, how they could no longer carry all of his great weight, how he could no longer navigate the slick and icy hills of his pasture in the winter. You have to go, I told him, and he saw it, that it was his destiny, his purpose.

I remember Simon the donkey showing me how much he had loved the poor farmer who starved him and left him for dead, how he set out to teach me what compassion was and what it meant. He showed me that he forgave the farmer for nearly killing him, and bore no grudges, did not know what it was to hate, and in this way he showed me the spiritual power of animals.

I remember Winston the II, a young rooster who nearly killed his gentle father; how I closed my eyes and visualized my sadness and how he showed me his rage and fury, and I knew I had to point the rifle at him and shoot him through the head. Harsh as it sounds, that is the real life of real animals. And the real life of the farm.

I talked with Rocky, our old blind pony, and he showed me the pain and fatigue and suffering in his old bones, and he told me very clearly that he no longer had the energy to deal with his daily life and was ready to go, and needed to go.

I recall Susie, our sweet old ewe, telling

me she was eager to give birth, healthy and ready to live with her offspring. So it was.

I saw that animals do not know or fear death as we do; they accept the limitations of their lives. They live as long as they can, they understand when it is time for them to leave, and they do so uncomplainingly and without regret or recrimination. This is the gift we can offer them, one we cannot often offer ourselves or our loved ones. We can help them leave the world in a loving and painless and dignified way.

If we would only open our eyes and learn, they would teach us how.

Animals are calling us to a new awakening, a new understanding of their rights and of ours. Instead of lobbying to banish animals from people, we must find ways to keep them. Instead of making it more difficult for animals to be with people — especially the poor, the working people, the elderly — we must make it easier.

Animals are a precious resource, like water and forests. They must never be banned without a clear understanding of where they will go, who will care for them, and who will pay for that care. Instead of intimidating, threatening, and harassing people who love animals and live with them we must find ways to help people keep them and care for them well.

Animal abuse is an awful thing, but the elimination of animals from the lives of people, and, eventually, from the earth itself, is worse.

Understanding animals also means understanding people, because animals cannot survive in the world without them.

I think of Tawni Angel in California, whose license to give pony rides to delighted children was taken from her because animal rights protestors decided it was torture for ponies to have to give rides to children. Two different police investigations found otherwise, but that didn't seem to matter. Tawni is struggling to survive and find a way to keep her ponies with her.

I think of Sarah L., a lonely eighty-two-year-old woman whose five feral cats were taken away from her suburban New Jersey home in a police van because secret informers reported they were living outside her trailer. The cats, she said, were her very life. She is lost without them.

I think of Harold, a draft horse put down in Southern California because his owner could no longer find work for the animal in movies. The film producers told him it was not worth the trouble of using animals in movies anymore. They were tired of fighting the accusations of cruelty and abuse.

I think of the farmer in Upstate New York, reported to the police by passersby who saw his cows standing out in the snow during a storm. Police seized his cows; his legal fees and boarding costs forced him to sell his farm. His cows were never returned.

I think of the homeless man in Manhattan whose dog was taken from him by the police and euthanized because a passerby saw an untreated sore on the dog's back. The dog had lived with his human for ten years. They were the whole world to one another.

The list grows and grows: the pony, the horses, the dogs and cats, one after the other, taken from the world, banned, killed, taken to preserves away from their people and lives and work.

Talking to animals means saving them. If they can tell us what they want and need, perhaps they can be spared the fate of having hapless and misguided human beings decide their fate. There is no more basic right for any animal than the right to survive.

One day in 2014, I decided to experience a carriage ride for myself and take my new friend, Ariel, up on his offer to treat me and my wife, Maria, to one of his famed midnight rides. We walked to Central Park just before midnight on a balmy evening in early spring. The night was clear, the temperature in the fifties. There were a handful of carriage horses lined up on the southern edge of the park.

Ariel was waiting for us. He is one of those people who love to give other people gifts and bring them pleasure; it just lights him up. He had invited Maria and me on a midnight carriage ride in part to thank me for writing in support of the carriage horses.

Ariel asked us if we were ready, made sure we had blankets, and then we headed out into the beautiful night. He handed us a bag of fruit, some bottles of water. I had been in the park many times, but Maria and I had never seen it like this, free of the rushing traffic, bicyclists, joggers, and pedicabs that pour through it every day.

The park was still. Rebecca's clip-clopping echoed off the big trees, their shadows enveloping us. The park was designed in large part for the horses. They have walked in it every day of its existence, since it was built in 1851. Olmsted insisted that the park be designed so that the horses and the people of the city could see one another, pass by one another. That night, I could see very clearly that the horses were the most natural thing in this beautiful space — it was the trucks and cars and pedicabs that were out of place, day or night.

Ariel took us through the park, past the big old streetlamps, to the beautiful fountains, and some hidden lakes and spaces we had never seen before. He stopped to light candles for us in the back of his cab. When we approached Strawberry Fields, an area of the park dedicated to the memory of John Lennon, who lived, and was killed, just across the street, a young man startled us by hopping into the carriage and playing Beethoven on his violin for the next twenty minutes. It turns out that Ariel had hired the young music student to entertain us.

The park is eerily beautiful after midnight, quiet and restful, so different than during the crowded days. The clip-clop of the horses is loud and distinct, the fountains a musical roar, the leaves rustling peacefully in the quiet.

Ariel knows all the nooks and crannies of the park. Once or twice, he stopped the carriage and walked us down a path and into the woods, the sky suddenly opening up onto a stunningly beautiful city skyline. As we were nearing the end of the ride, Ariel turned toward Maria and me and tossed rose petals onto our laps. He then sang several songs he had written.

The two hours we spent riding through the park was a wonderful token of gratitude for my support, but the message behind the carriage ride was much more powerful. Ariel offered us a trip into the past; more than just a ride through the park, the experience evoked the thousands of years that horses and human beings have been living and working together.

For me, the ride confirmed that horses belong in New York, that it's their park, too.

Whenever I contemplate what a new and more mystical understanding of animals could look like, I always think of the horses and of that magical midnight ride. Any true understanding of spirituality means thinking deeply about what we mean by nature, what we mean by living with animals. Just as peace means much more than the absence of war, treating animals well means so much more than hiding them away from people on rescue farms.

Spirituality is reflected in a lifestyle that is balanced, that leaves us with a capacity for love and wonder. The carriage horses speak to us of love and wonder, but how can we hear them amid constant noise, interminable and jarring distractions, in a life bounded by greed and worry and political conflict?

A balanced life is a life in harmony with nature, with animals, with inner peace and strength. That is what I felt in Ariel's carriage, what I feel when I stand with the horses, when I visit them in the stables. It is an attitude of the heart, one that appreciates animals with a serene and open attentiveness, that is capable of being fully present. Jesus invited human beings to contemplate the lilies of the field and the birds in the air; animals like the carriage horses can show us how to overcome the anxiety and alarm that plague us and can make our world angry, superficial, aggressive, and wasteful. We love animals because they offer us love and affection, they can't argue with or contradict us, they don't care what we wear or look like, they are reliably loyal and affectionate. In our fragmented and contentious world, that is important work. In fact, it is the new work of dogs.

It was a great gift to see that my love for journalism still lived inside me and could now be applied to my other love — communicating with animals. These disparate worlds of mine — my past, my present, my passions, my blog, my farm — all suddenly came together. I could tell the stories were having an impact, I started to get large numbers of messages saying, "Thanks for making me think." Sweet noise to any writer.

One day I got a telephone call from Pamela Rickenbach, the cofounder of Blue Star Equiculture, a draft horse sanctuary and retirement home in Palmer, Massachusetts. I had talked to Pamela about the carriage horses and how they were speaking to me, urging me on in writing their story, and we had become friends. I told her I was mystified by the horses' messages. Pamela told me she had shared my writing with a much loved and respected Sioux spiritual leader, Chief Arvol Looking Horse. He was, she said, the Dalai Lama of the Sioux Nation. Chief Arvol rarely asked to meet people outside of his own world, Pamela said, excited, but he had read my work in support of the horses and wished to meet with me when he came to Central Park that summer to speak at the United Nations on behalf of world peace. I was flattered and humbled.

So, Maria and I went to New York City on a warm summer day to meet with the spiritual leader of the Horse Nation.

Chief Arvol, 19th Generation Keeper of the Sacred White Buffalo Calf Pipe Bundle, looked the part of a Native American chief.

I could easily envision his image etched on the rocky façade of a mountain. He was imposing, tall, and handsome, with long braids and a weathered face. He wore a beautiful and colorful embroidered shirt.

We ordered sandwiches and falafel from a nearby stand and went to find a place to sit. Chief Arvol is so charismatic a figure that people in the park stopped to stare at us; some even sat down in a circle around us, to watch and listen.

Chief Arvol and I talked for nearly an hour. He told me that the horses were indeed speaking to me, that they had prayed for me to tell their story. In the Native American culture, he said, people spoke with horses all the time. Their culture respected the sacred and ancient bond between people and animals.

In Western culture, he said, that bond had been forgotten; people had lost respect for the partnership between people and animals and were driving the animals away.

It's hard for me to understand how the horses can be guiding me, I said.

You don't have to understand, he said, you just have to accept it. And you *are* accepting it, he said. I can see it in your writing.

I took the Chief's advice. I stopped resisting the idea that the horses were talking to me, and just accepted it. I kept writing.

To be honest, I could not have stopped if I wanted to. It seemed that this was a calling for me, a chance to expand my writing about animals and communicating with them in the most important way and for the best reasons; a chance, too, to call upon the investigative and reporting skills I had learned working at some of the best newspapers in America, a chance to use my blog platform for a good cause; a chance, I believed, to try to help correct a great injustice.

Chief Arvol said the fate of the carriage horses was more important than the people in New York realized. The horses were the symbols and representatives of Mother Earth. If they were taken from the city, he warned, they would take the wind and rain and fire with them. The world was at a crossroads, he said. People would either learn to live in harmony or perish together on a desecrated planet.

I believed him. After that day in Central Park, I had no doubt about it.

I honestly don't know what impact my writing had on the struggle to save the horses in New York City, but I do think I was able to give the disheartened drivers some hope and support. When I go to New York City and walk in the park, the drivers always greet me, shout out to me, thank me. I have received thousands of messages from people all over the country, saying they view the horses in a different way, that they now understand there are many complex issues surrounding the future of animals in our world.

I believe I've helped some people understand that all animals are not pets; they are not fur babies. Domesticated working animals need to work. Saving animals does not mean confining them to often impoverished and struggling rescue farms. It means understanding their true natures and helping to keep them in our everyday world, rather than taking them away from people, who will otherwise never lay eyes upon them again.

As it happened, the mayor and the animal rights organizations seeking to ban the horses failed. There was overwhelming opposition to the notion that the horses were being abused or should be banned. The mayor did not ban the horses on "day one," or in his first week, or even after day five hundred. At the end of a long and bitter effort over a year and a half to rally support in the City Council and the public for banning the horses, it was clear the mayor and his allies had failed miserably.

There were not even enough votes to get the ban-the-horses bill out of the City Council's Transportation Committee. A second effort to resolve the carriage trade conflict also failed in the winter of 2016, when the mayor unsuccessfully introduced legislation to restrict the carriage trade and move the horses into a renovated building in Central Park.

But this is not a happy ending, and there probably isn't one. Carriage drivers continue to be targeted and mercilessly harassed. Tourists who seek a carriage ride and their children are often taunted, called murderers and Nazis. It is not uncommon for parents to flee the carriages, rushing away with crying children who have been accused of murdering animals. The mayor still promises to ban the horses as soon as he can figure out how to do it.

The forces threatening the horses are enormous. They include the mayor, real estate developers, and fanatic animal rights activists who seem beyond reason or learning or empathy. Many people in the carriage trade have already decided to get out. The carriage trade no longer represents the American Dream for them or their children.

So many of the immigrants who work in the carriage trade came to America to look for lives of freedom and meaning, and sadly, they see no long-term future now for themselves or their families. The city could so easily commit to keeping them there, but instead, the government seems committed to driving them away, and perhaps Chief Arvol's fearful prophecy will come true in New York.

I think he is right. Our Western culture has forgotten the long and precious history that people and animals have. If the horses leave, they will take the wind and rain with them. And much of the magic.

I thank the horses. They gave me a wiser and more mystical understanding of animals than I ever had, or ever imagined having. As Pope Francis has suggested, we are at a crossroads when it comes to animals. We can keep pushing them out of our cities and populated communities under the guise of protecting them from abuse, or we can begin understanding their true needs and look for ways to keep them among us. We can turn their fates and futures over to people who seem to know nothing about them, or we can reclaim the right to speak for the real needs of real animals in the real world. We can continue to use animals as a battering ram against human beings, or we can learn to respect people and animals both and help them to life safely and well together.

When someone asks me what the carriage horse controversy is truly about, I say it isn't about real estate or animal welfare or traffic safety. It's about an attitude of the heart. The animals need us. Their most elemental right is the right to survive on the earth, and our most elemental task is to understand them well enough to know how to make that happen.

If we ask them, they will tell us.

9
JOSHUA ROCKWOOD

In March 2015, the northeastern United States, and its many farms, were hit with one of the worst cold spells in history. In some places temperatures plunged to as low as minus-30 degrees at night. Tens of thousands of water pipes, sewer lines, feeding tanks, streams, and brooks froze over.

It was an awful time, especially for farmers, and it was a very hard time for Bedlam Farm. The frost-free water line we had dug the previous autumn — it was five feet deep — froze in late February. Our donkeys and sheep were confined to the Pole Barn, the drifts and ice blocked them in. They had shelter, but nowhere to walk or forage. The ground was covered in several feet of rock-hard ice and snow. We could not get to the hay feeders for weeks, and had to dump the hay in the barn and sometimes on the frozen ground.

We hauled buckets from the farmhouse to

the heated water tank outside two or three times a day, and sometimes, when the power went out, or even if it didn't, the tanks froze and the animals had to wait a few hours before we could get water to them. They had to drink it in a hurry, and they did.

We checked each day to make sure they were healthy and as comfortable as possible. They did fine. We reminded ourselves that these conditions were common in nature. Animals know how to deal with it better than we do.

But things did not go so smoothly for a young and idealistic farmer named Joshua Rockwood, who leases a ninety-acre farm in Glenville, New York, where he raises cattle, pigs, sheep, and chickens. He lives right down the road from his farm with his wife and two young children. Joshua's mission is to raise healthy meat on the farm and sell it to local people.

His farm is called West Wind Acres. His cows are usually up on the hill, and there are freshwater streams running year round. The pigs and sheep graze on a big, wide pasture. The chickens have a stylish two-story henhouse in a large open field; they too are free-range. The Maremma guard dogs live with the sheep. In the barn, there are huge pens for dogs and for the goats when the animals need to shelter inside.

Joshua is a sort of farm geek. He's obsessed with nutrition, and feed, and aspires to sell the best meat, raised in the healthiest and most environmentally sound way. His business is new, but he's attracting a growing list of customers happy with the quality of his food. He has many detailed plans to expand his business in the region.

West Wind Acres is very much a newcomer's farm. It is a patchwork of tin huts, homemade shelters, and trailers; equipment, water pans, and buckets are strewn everywhere. Joshua is the first to say he is inexperienced, that he is learning as he goes. He also points out — correctly — that all of his animals are very well cared for: in good (healthy) coat, alert, active, healthy.

Joshua doesn't have much room in the outbuildings of his farm, so he keeps hay and food down the road near his house. He was not prepared for February temperatures that plunged so far below zero. Neither was I. Neither were any of the farmers I know.

One night, someone — he will probably never know who — drove by Joshua's farm and saw his Maremma sheepdogs running loose in the field, as guard dogs do. This person called the police and said they were concerned about the well-being of the dogs, running around in the cold.

Maremmas are famed sheepdogs who, in all weather, live outside with the sheep, fending off coyotes, wolves, and stray dogs. They have thick, curly white coats, and suffer greatly when confined indoors. Joshua also had two draft horses and a pony; their hooves, he knew, were overgrown by about two months. In the bitter cold, many farriers skipped their rounds for a few weeks, or even a month or two. It does not harm horses for their hooves to be overgrown for a few months.

The police notified the local Humane Society and also contacted a local horse rescue farm. They came by Joshua's house and said they wanted to see his farm; they were seeking a warrant to inspect his animals. Joshua was shocked, but happy to cooperate with the police. He answered their questions about his water tanks and whether or not they were frozen.

When the police left, Joshua, concerned, called two local vets he had worked with and asked them to come out to the farm and examine all of his animals. They did. Both vets found the animals to be healthy, hydrated, and well cared for. They each signed reports testifying to that.

In a day or so, the police returned. With

them were an officer from the local Humane Society and two trailers from two different rescue farms.

Joshua had been late getting to his farm that morning. He had taken his son to see his grandmother. He knew that his water buckets had frozen, as they had on almost every farm in the Northeast, and that the streams in the hill where his cattle drank had been covered over with ice. Each day, he had chipped holes in the ice cover and brought water from the farmhouse for the dogs and sheep and pigs.

None of his animals had died, or become dehydrated. His pigs did what pigs do: they gathered up in huddles and burrowed into hay and compost. All the animals slept under shelters with roofs.

The officers asked Joshua where the hay and feed were for the pigs and sheep. He said they were stored at his house, that he had no room in the barn. He told the officers he made sure there was fresh water each day. The officers found animal feces frozen into the water bowls and buckets. Joshua explained to them that animals had poor toilet hygiene and that he cleaned the buckets out regularly.

The officers examined the hooves of the horses and conferred with the horse rescue people. They seized all three of his horses and ordered them shipped to a horse rescue farm.

By the time the police left that morning, Joshua had not only had his horses seized, he also had been cited thirteen times for animal cruelty. The charges included having an unheated barn, horses with overgrown hooves, frozen water receptacles, and two pigs with gray matter on the tips of one ear that might have been frostbite.

The police asked a judge to set bail for Joshua. They said they considered him a flight risk, even with a ninety-acre farm, more than one hundred animals, and a wife and two kids.

The police refused to say who had informed on him. Under the law in New York and many other states, that information can be kept permanently confidential in an animal cruelty investigation. The idea is to make it easy and comfortable to report animal abuse, without fear of confrontation, retaliation, or challenge. It also, it seems, makes it easy and comfortable to inform on neighbors and strangers and farmers without any kind of accountability.

I know a lot of farmers. Almost any small farmer will be happy to talk to you about informers. Farming was once one of those difficult but obsessive ways of life that celebrated individuality, freedom, and privacy. Nowadays, though, almost any farmer you meet will have stories about the police or sheriff pulling up because somebody drove by and saw a horse or cow lying down, or a cow out in the pasture with some snow on her back, or a horse pulling a wagon on a warm day, or a border collie running alongside sheep with his tongue hanging out.

Those were once considered common, even beautiful sights on a farm. City people loved to ride out into the country to see such scenes.

Today, the context of farming has changed. Small farms are struggling, unable to compete with the giant corporate industrial farms. And those iconic scenes of animals working and living outside are often reasons to call the police and have animals seized. The farmer is forced to try to explain why work is not abuse for big horses, and working dogs need to work and run, and cows with their thick hides have no problem standing out in the snow to enjoy some fresh air.

The farmers live with their animals and know them. People driving by with their cell phones at night and calling the police usually do not.

This is why nearly three hundred farmers showed up to support Joshua Rockwood at his first — and second and third — court hearing. It could have been me, they said to themselves, and quite often it was.

As I stood in line with the farmers waiting to be searched by the police before entering the courtroom, I heard story after story. Abuse seems to be undergoing a radical redefinition. The farmers all say that the people who used to regulate them came from farming families, grew up in rural areas with animals, and knew animals.

They all report that this no longer true. The people regulating animals now are ideologues and activists, who seem to know nothing about farming or the real needs of animals.

What is important about Joshua Rockwood's case isn't that he is a saint or perfect; it's that his arrest highlights the fearful new reality for animals in America. It is becoming increasingly difficult, fraught, and expensive to adopt or acquire them, or to live with them in peace and privacy.

The police found a healthy working farm when they arrived at West Wind Acres. There were no sick or dying animals who had to be rushed to treatment. At the same time that Joshua's water tanks froze, the town hall's sewage pipes froze, spilling waste onto the floors of the municipal building. No one was arrested.

It was moving to see the outrage and outpouring of support for Joshua — farmers sent money via crowdsourcing sites, took precious time away from work to attend his hearings, sent letters and messages of support. Farmers from all over the country told Joshua the same thing had happened to them. Many had been put out of business due to legal fees; others settled because they couldn't afford the long and costly process of a trial. Still others were publicly shamed; many lost their farms.

I didn't get to see their farms, so I can't say what condition their animals were in, but I did get to see Joshua Rockwood's, and I can tell you his animals were healthy, alert, and well cared for.

As I mentioned earlier, abused or starved animals are easy to spot — they are sluggish, wary, weak. Their eyes are runny or rheumy, their ears back, their movements unsteady, unsure. No animal in Joshua's care displayed these signs.

Joshua's farm is not a pristine vision out of a children's picture book. His animals huddle under aluminum sheds, wooden cabins, and one big barn. Farm shelters often look liked abandoned refugee camps to the uninitiated, but they do what they are supposed to: they shelter the animals from the elements. They do the job. But one of the curious anomalies about shelter arrests is that the people arresting the farmers tend to believe that animals ought to live in the wild, in nature. They seem to forget that animals in the wild do not have shelters. They gather under trees and bushes in storms and bad weather; they huddle together in the cold.

Few animals anywhere have heated barns. In fact, domesticated animals like sheep and donkeys and cows dislike heated barns; they are bred to live outside all year round. In nature, they suffer from bugs, heat, cold, mud, predators, the scarcity of food, and battles with one another. Joshua's animals, however rickety his shelters, are treated better than any animal in the wild has ever been treated.

Among the accusations made against Joshua was a citation for having inadequate shelters because an officer saw patches of gray on the ears of two pigs. The police consulted with a vet who said she thought it might be frostbite.

Joshua has more than one hundred pigs at

any given time; he makes sure they are fat and healthy because he sells the meat to discriminating customers who want to know where the pigs come from and how they are fed and treated. He is admirably transparent. His blog, Westwindacres.com, provides details on the feeding and care of all of his animals, and the nutritional values of the meat he sells.

A small army of pig farmers showed up at his first court hearing. None of them had expensive shelters. Pigs, they told me, huddle together with one another and crawl into compost and muck to stay warm.

"When it's minus-twenty-seven degrees," Stan, a pig farmer from Massachusetts, told me, "some will get frostbite anywhere, even inside of a barn. Usually it's the tip of an ear. They don't feel it much there, and it doesn't affect their health."

This has been my experience, too. I've had sheep get frostbite on the tip of an ear when the windchill brought the temperatures down to minus-30. The sheep were inside a well-built cow barn with thick windows and doors. A month or so later, almost all of the ear had grown back. The ewe lived for a number of years. She had four babies, and she ran so fast she gave the border collies fits.

Why were the farmers so upset? Almost all of them are animal lovers; they live with animals every day. They have no patience for abuse or animal cruelty. But the new and sometimes hysterical obsession with animal abuse and cruelty is criminalizing farming.

Few farmers have money for big new fences and fancy shelters; they always patch things together with whatever they have on hand. Farms can look like bomb sites, their yards and barns filled with junk. Farmers are resourceful — they throw nothing away, and they can always recycle some leftover piece of scrap wood or metal. They live by their milk checks or corn and alfalfa crop, and no two years are alike. They always have to be prepared for a drought, or a long heat wave, or a brutal winter, or rainstorms and flood. Their lives can be upended at any moment. They fight government regulators, bureaucrats who control much of their lives, animal rights activists who hound them with informers, and ridiculous laws that cost them money and do animals little good.

When the police come and raid their farms and take some of their animals away, almost always based on the whispers of secret informers, and their photos are put on television, it is very often more — much more — than they can afford or absorb.

Joshua was just getting started with his business, beginning to build up a client base that was signing up for shares in his farm, that wanted his poultry, beef, pork, eggs, and lamb. He was signing up outlets and potential distributors, figuring out rotational grazing, nutrition, and water. Here is a hardworking farmer whose animals are healthier and happier than any one of the nine billion animals on industrial factory farms. None of his animals had died; none were sick or injured. Two pigs out of one hundred had a sliver of gray tissue on their ears and his horses' hooves were slightly overgrown, a couple of months behind their regular trimming time.

Touring Joshua's farm shortly after his arrest, I grew angry at the people and forces who were trying destroy yet another farmer. Don't these people know anything about farms? Don't they know that streams freeze outdoors all the time and the animals eat snow — there was plenty on the ground — or poke around the edges of the ice to get water? In Joshua's case, his animals didn't have to wait very long since he came out every day to break the ice and fill the bowls with water.

This is farm life, and the police, many of

whom have never set foot on a farm, are being drawn into the middle of the deepening social conflict between the animal rights movement, farmers, and many animal lovers.

Joshua said it was very clear to him that the police who came to his farm had no real idea how animals lived or were treated on farms. They deferred to the Humane Society officer, who seemed to find violations everywhere, and a small animal vet and some horse rescue people who came in two trailers, obviously prepared to take his horses away.

It's also worth nothing that when this happens, the people whose animals are taken away often have to pay staggering boarding and administrative fees to get them back, whether they are found guilty or not. In Joshua's case, he was asked to pay $28,000 for one month of boarding and veterinary fees for his three horses, long before his trial even began.

Almost everyone in his community saw photos of Joshua as he was fingerprinted and booked in the town police department, and his mug shot was all over local television, along with some footage of his horses' hooves. He lost many customers right away. Who wants to buy meat from a farmer who has been branded as an animal abuser? Who would take the time to question the claims or hear both sides of the story? Joshua's customers were not given that opportunity wince the news media did not report that two different vets found the animals at West Wind Acres to be healthy and hydrated. It seems these findings will never make the news.

Joshua is determined to survive, and to keep his farm going and fulfill his dream of selling the healthiest possible food to his neighbors and local people. It's going to be hard. As of this writing, Joshua and his lawyer had been offered a deal by the prosecutor. All of the charges but one were dismissed, and the last will be stricken from the record by the end of 2016. In effect, Joshua was exonerated. A columnist for the *Albany Times-Union* wrote that Joshua should never have been arrested.

Joshua raised $8,000 online in order to get his horses back. Joshua will never know who the secret informer was who nearly ruined his life or have an opportunity to confront him or her in court.

Joshua had been offered a plea deal several times. He refused, saying he would never plead guilty to a crime he didn't commit. Farmers and animal lovers raised more than

$70,000 online to help Joshua fight a case that seemed patently unjust. Without them, he would almost surely have lost his farm and perhaps also his court battle.

Several things about the police case against Joshua were troubling to me. If Joshua's animals were treated so cruelly and abusively as to warrant thirteen separate charges, why drop almost all of them without even going to trial? Also, after the raids and seizure of Joshua's animals, no police officer, horse rescue person, or Humane Society official ever returned to the farm. Not even once. Nobody came back to see if there was enough water, feed, and shelter.

If the concern about the animals was serious enough to warrant disrupting someone's entire livelihood and life, then why not come back even once to see how the animals were doing? And if the farm was so inhospitable as to warrant thirteen different charges, why leave animals there at all? Why not remove all of them to safety or better care?

Through no fault of his own, Joshua Rockwood was drawn into an Orwellian nightmare of informers, dogma, and propaganda disseminated by a lazy and pliant media. It seems shortsighted for the animal rights movement to target farmers rather than to reach out to them and help support their work.

Like the carriage drivers in New York, farmers are essential to the future of animals and their survival in a changing world. The animal rights movement — and the police and legislators who have become entangled with the movement and its lobbying — have turned farmers into enemies. In fact, farmers are the best hope of survival for many animals in our world.

10
"Good Morning, Equines"

Every morning on our farm, after we wake up, call for the dogs, put our boots on, and go outside, Maria turns to the pasture gate, where there is always a crowd waiting for her, and yells out, "Good morning, Equines!"

We are greeted with a chorus of whinnies and braying.

The new day begins.

When I first knew Maria, before she became my girlfriend and then my wife, we had an arrangement. In exchange for using one of my barns as a studio for her fiber art, she came to the farm on weekends to help take care of my sheep, donkeys, chickens, and barn cats.

Maria's job was to make sure the animals had water and fresh feed and were all healthy and moving around well. She came twice a day, once in the morning, once at night. She would park her small Toyota at the base of the driveway, walk up to the gate, and let herself into the barn and feeding area.

At the time, I wrote in a small study that looked out over the pasture, so I could see the animals. I had this fantasy of having an office with a Dutch door — shades of *Mr. Ed* — so my donkeys, Lulu and Fanny, could stick their heads in from time to time, to get a carrot or a scratch on the nose, but that didn't work out.

One morning, I heard Maria's car pull in and saw her walk up the driveway. She waved to me (my office was close to the driveway) and then she opened the gate and walked in. Maria's pockets were always stuffed with apples and carrots. The "equines," as she called the donkeys, loved her from the first and came trotting over to her to sniff her pockets and let her brush them.

One of the first thing I learned from Maria about donkeys and horses is that they love attention; they love to be tended to, just like people.

And Maria was to teach me something else — equines love to talk.

As I watched, I saw Maria kneel down on the ground. Her head was just below the level of the donkeys' heads. She handed out a few carrot chunks, and the donkeys stood close to her, munching their carrots.

I expected her to get up and bring out the hay, but she didn't. She knelt before the donkeys, quiet and low to the ground. She closed her eyes, and Fanny and Lulu leaned in close to her.

The three of them stood that way for the longest time. I was nearly hypnotized by their intense connection to one another. *They are talking,* I thought. *They are having a conversation.* Maria was listening to them, and they were listening to her, their three heads almost touching. There was a shimmer between them, a peacefulness.

I went outside and waited for her to come out of the barn. I had my camera, hoping to capture the interaction between Maria and the donkeys.

"You seemed to be talking to them," I said.

She nodded, flushed with excitement. "It felt like I was."

"What did they say to you?" I asked.

"It wasn't really in words," she said, "more feelings. They felt close to one another, to me. It felt as if the three of us were one thing. It felt as if they understood who I was, who I am."

That was all she could say about it that

day, although her dialogue with these two sweet and intuitive creatures was to deepen beyond either of our expectations.

I started taking pictures of Maria with the donkeys, started watching closely. It was similar to my dialogue with Elvis, but also quite different. The donkeys seemed more evolved than the big Swiss steer, more intuitive, more open and naturally affectionate.

Maria was going deeper than I had. The animals had transmitted to her a feeling of peacefulness, of perspective and well-being. A healing feeling that she badly needed, as she was going through a painful divorce at the time. And she received a very clear sense that they trusted her.

Maria said she had cleared her mind. She felt a powerful and very healing kind of quiet, something she was receiving from them. A welcome, an acceptance, a purging of unnecessary words and assumptions. She was listening to them.

Six years later, we were married and living on our farm in Cambridge, New York, the second Bedlam Farm. When we moved there, we inherited Rocky, the blind old Appaloosa pony who had lived out in the pasture for fifteen years by himself. There were few fences, no shelter, no heated water tanks. Rocky fended for himself, and he did very well.

There was something both touching and heroic about Rocky. He seemed to endure every hardship with grace and instinct. He had made his own trails in the snow and mud and had his secret paths back to the streams behind the farmhouse.

Maria was especially drawn to Rocky. She instantly found ways to communicate with him. She would enter the pasture with apples and carrots and speak in a clear but soft and soothing voice. She would stand still and wait for him to locate her so he wouldn't be anxious about where she was or startled when she came up to him. She would touch him, slowly and carefully, and scratch his neck.

For the first few visits, she always had a treat, and she always offered it while speaking to him, explaining who she was and what she was doing.

Rocky's owner was more than one hundred years old when she died, and Rocky was alone for many years before we came to the farm. Like most equines, he loved attention.

When Rocky heard our car, he would whinny and come up to the fence. Maria would talk to him as she approached, and so would I. We always had apples or carrots for him, and he seemed to love being near us. He bonded closely with Red, my border collie, who started instinctively doing seeing-eye work with him, leading him down the paths the pony walked and steering him and around obstacles.

Rocky's coat, mane, and tail were rough and matted, so Maria began brushing him. He was skittish at first, because he was blind, I think, and alone for so long. Quite often, Rocky would back away and trot down the hill to put some distance between himself and us when he sensed we were going to groom him. Maria was gentle, patient, persistent.

After a few weeks, Rocky got comfortable. Maria would lean down in front of him, let him sense her, feel her emotions. She would imagine herself grooming him, combing out his tail, walking with him around the pasture. It all came to pass. In a month or two, he was her horse, and she loved him deeply.

Maria intuitively did many of the things it took me years to learn and feel. She is an emotional person, and her emotions live close to the surface, where animals can read them easily.

Women are often more open with their emotions than men. But access to emotions is critical in talking to animals because animals are so skilled at picking up on them.

It is inspiring and instructive to watch Maria communicate with her "equines." She never patronizes them, emotionalizes them, or sees them only as pitiable creatures for her to save. She sees them as partners, companions, fellow dwellers with us on the farm.

When speaking with Rocky or the donkeys, she clears her head naturally of all the words and narratives people love to put in the heads of animals — narratives that block understanding. And she clears her head of anxiety, anger, or frustration. She approaches the animals with an open heart, her emotions available to them. She uses food skillfully to get their attention, earn their trust, open up the channels of communication.

Maria's agenda, when she has one, is clear, simple, and easily discernable to the animals: she wants nothing from them but their attention — if they wish to give it — and an opportunity to exchange feelings and emotions with them. She never personalizes her communications. If the equines are distracted by fresh grass, or deer in the field, or the sexual vibrations and emotions of one another, and don't wish to talk to her, she accepts that without complaint. She never tries to manipulate them into loving her or talking to her.

She is an active listener, a rare thing for humans in the animal world who claim they know what animals are thinking, and struggle to embrace the humility of accepting that we rarely do. Maria does not presume to understand their language or force words and thoughts onto their consciousness from the human language. She marvels in their alienness and respects it. She does not see them as brothers and sisters; she does not see them as dependents, as inferior or superior; she does not seek to impose human rights and values on these alien creatures. She knows it makes them uncomfortable, and she knows it is rarely in their true interests.

She communicates viscerally, imagining the relationship she wishes to have with them, the closeness, the quiet, the listening.

Maria brings an almost perfect package of skills and instincts and emotions to talking to her horse and donkeys. If she has a slight weakness, it might be that she is not fully confident in her abilities. Sometimes she has a hard time seeing how much these animals love and trust her. Sometimes she has to remind herself, or, as often happens, the animals need to remind her. And they do. They bray at her, whinny at her, rush to her, stand by her to await her grooming and brushing.

These are the threads and strands of communication, this listening and feeling and sharing of emotion, this use of food as the glue of trust and attention.

Rocky's story took a sad turn when we brought our donkeys to the farm and they could not and would not accept a blind old pony into their herd. Equines in nature assault and drive off injured, old, and impaired horses. They are a grave threat in the wild because they attract predators and endanger all the other horses.

Simon, our male donkey, tried everything to drive Rocky away. He rammed the old blind pony, bit and kicked him, tried to drive him through fences and out of the farm.

Rocky was old, and greatly unnerved by these continuous and frightening assaults. He had no way of protecting himself other than to hide in the farthest corners of the pasture. For months we tried to separate him and Simon, keep them in separate pastures, give them time to get to know one another, but nothing worked. Rocky was acutely aware of Simon, and animals do not trust fences as much as people do. To Rocky, Simon was always there, always near, even if he couldn't get to him. He was always aware of Simon. He could hear him, smell him, sense his rage and hostility.

Rocky began to weaken and fail, and our vet concurred that it was best to spare him another winter full of tension and struggle. So we put him down.

Maria was shattered by Rocky's death. She and the sweet, old pony had became so close. They talked all the time. I loved photographing the two of them together, out in the pasture, Rocky soaking up the attention, the affection, the soothing voice and songs through which Maria communicated.

When you practice communicating with animals, one thing leads to another, one experience opens the door to another, one learned thing can be carried over to another.

As a result of our becoming involved in the campaign to save the New York City carriage horses, we were invited to Blue Star Equiculture, the ninety-acre farm in central Massachusetts I mentioned earlier. Blue Star was founded in April 2009 as a sanctuary and retirement home for the carriage horses. Two former Philadelphia carriage horse drivers, Christina Hansen (now a carriage driver in New York) and Pamela Rickenbach, are responsible for creating and funding this haven.

When Maria and I went to Blue Star, we found the horses had entered our heads and shaped our lives once more. The draft horses at Blue Star are enormous and beautiful to behold.

Since Blue Star was founded, Hansen moved to New York City to drive horse carriages there and to help in the fight to keep the horses from being banned by animal rights activists. Pamela and her late husband, Paul Moshimer, took over the farm and ran it in ways that greatly transcended the original mandate. It has become an organic farming center and workhorse rescue farm. At any given time, there are about thirty-two working horses at Blue Star. The farm is always looking to adopt the horses out for healthy and meaningful work, the kind draft horses have been doing for thousands of years.

Pamela Rickenbach is a fascinating person, a kind of horse historian, mystic, and spiritualist. From the first, I was struck by her vision for the future of animals in our world, and her amazing gifts when it comes to understanding them and communicating with them.

Pamela believes in the sanctity and value

of the horses' work with people. She believes animals should never be used to harm or hurt people, but to support them and share life on the earth with them. She believes, as I do, that it is not possible to love animals and hate people, and that the future of animals depends on the willingness of people to live and work with them, and on the support people are given to do so. There is a sharp and telling contrast between the tensions and rigid ideology of the animal rights movement and the powerfully humane vision of Blue Star:

Animals should not be removed from the everyday lives of humans, but kept among us.

Animals face near complete extinction on the earth; it is an environmental imperative that they be given opportunities to once again work with people and live among people.

The people who live and work with animals need support and understanding, just as the animals do.

I believe — and Blue Star preaches — this powerful idea: the more support people have, the better care the animals have. There is no better or quicker or more efficient way to help animals than to support farmers and carriage drivers and, yes, elephant trainers and pony ride operators, and the poor and elderly animal lovers of the world, than to offer them encouragement, assistance, and affection.

When she met the big horses of Blue Star, Maria's connection to horses and other animals deepened. Building on her experiences with Rocky and the carriage horses, she began regularly visiting Blue Star and volunteering with the animals.

Despite these powerful and immersive experiences, the idea of bringing a horse to our farm was not something we seriously considered. But all that changed as a result of the death of our beloved donkey, Simon.

One morning, Maria and I got up and I looked out of the bedroom window that faces our pasture. Usually, the animals are all standing at the gate, waiting for us to wake up and bring them hay or apples or carrots. That morning, as I looked out the window, I saw Simon shake his head suddenly and severely. He looked confused, off-balance. He was always the first one at the gate, always listening for us. He loved his food.

"Something is wrong," I called out to Maria. "I think Simon had a stroke."

We rushed out to the pasture. I took a video of Simon with my iPhone and emailed it to Ken Norman, our farrier. I also sent it to Granville Veterinary Service, our large-animal vets.

Ken texted back me in seconds. He was in the hospital at the time, recovering from knee replacement surgery. Doesn't look good, he said. Call the vet.

Dr. John McDermott, our vet, was already on the way.

Simon was getting wobbly on his feet and we helped guide him to the ground. His pupils were dilated, and he seemed completely disoriented.

Maria and I lay down next to him. I closed my eyes and laid my head against him. We had been through so much together. He was so loving and grateful. He had suffered so much.

Maria said she received what she thought was an image from Simon, a good-bye, an image of an open doorway. Then Dr. McDermott arrived. So did Deb Foster, our farm and pet sitter.

The vet examined Simon carefully, listened to his heart, checked his eyes. "He's dying," he said. "His heart has almost stopped." Simon's breathing was getting shallower; his eyes were closing.

We agreed to ease Simon's passing. Dr. McDermott gave him an injection to stop his heart. Simon's suffering would soon be over.

I closed my eyes and imagined a peaceful and quick passing for him. My mind flashed back to the night in the pasture when Simon first came to our farm, nearly dead from neglect and starvation. I had put his head in my lap and hand-fed him fresh grass. Now I felt his warm breath on my face. I had an image of happiness and release, of the next thing, even as I knew that Simon's eyes were closed and he was gone.

In the country, when there is trouble, driveways and homes fill up with people. Ken Norman's wife, Eli Anita-Norman, came to sit with us in the pasture with Simon's body. So did several of our friends and neighbors. We never figured out how everybody knew, but that is also the way of the country. Everybody knows.

Dr. McDermott urged us to drag the body behind a tree. A secret informer of the "animal police" would surely call the authorities if they saw a donkey lying in the pasture.

"You don't need the police coming here

at time like this," the vet said.

"But it's nobody's business," I replied.

"Unfortunately, it is."

We dragged Simon behind a hill near the hay feeder. A half hour later, a tractor pulled into our yard. Vince Vecchio, a neighbor, unloaded his tractor from the trailer, and asked us where we wanted Simon buried.

Vince dug a hole in the ground, and with Deb's help we pulled Simon into it. We said our good-byes there.

Later, in the living room, I saw Maria and Eli connect with one another. Maria was upset over Simon's death; Eli was worried about Ken's surgery. She was also struggling to care for the more than thirty horses on their farm, many of them rescues. It was hard with Ken in the hospital.

We said we would drive up and visit her soon. She said that would be fine, and that there was a pony Maria might like to meet. Her name was Chloe. She was rescued from a farm whose owners could no longer care for her — a common story in the horse world.

A week later, Maria was standing in Eli's pasture staring at a Haflinger-Welsh pony who seemed calm, affable, and alert to both of us. I could see Maria wanted to get a horse. I strongly encouraged her to do it. It was time, and it seemed to be the most natural flow, a continuum — Rocky, the carriage horses, Simon, and, now, Chloe.

Maria and I had both received messages from Simon. They were about acceptance of life and moving on, opening doors, continuing our work with animals. Two months later, Ken was walking around without any pain, Eli and Maria had become close friends, and Maria was taking riding lessons.

It was not long before Ken and Eli pulled into the farm in a big trailer and walked Chloe out and through the pasture gate. The donkeys were wide-eyed but curious. The sheep fled to the outer pasture. Red, my savvy border collie, pretended not to even see the new arrival. Chloe meant nothing to him.

It was not a new thing, but the old thing, a continuation of the work, of our lives. That was Simon's message to both of us. Continue, life goes on. Finish the work.

Before Chloe came to the farm, Maria had kneeled down with the donkeys and told them the pony was coming. She used words, feelings, images. One morning, she said she heard a very clear message from Lulu and Fanny. A phrase came right into her head: "same thing."

The donkeys were telling her, she said, that it was no big deal to have a horse, that it was pretty much the same thing as another donkey. It was not something to worry about, not a problem for them.

We were skeptical of this message. People had told us the same thing when we brought Simon to be with Rocky on the farm, and it wasn't okay, it was a nightmare. But weeks later, when Chloe came, the donkeys' message turned out to be prophetic. It was no big deal.

Chloe came into the pasture, chased the sheep around for a while, and then joined our little equine herd. It was the same thing: the donkeys accepted her and she accepted them.

We had learned a lot from Rocky; if you can talk with an old blind pony, you can certainly figure out how to talk to a young and healthy one, right?

Working with Chloe was far more challenging for Maria than working with donkeys. Our donkeys are strong-willed, as donkeys are, but they are calm and affectionate. They are always willing to talk, and they are always listening. They know my mood the second I come out of the farmhouse. If I am grumpy, they are gone; if I am happy, they wait, braying softly, hoping for a carrot.

Ponies, we have learned, are a very different kind of animal. They are notoriously headstrong and independent. Although many people think of them as small and cute, that is not really their history or lineage. They were originally bred as work- and warhorses. They are strong and brave and difficult. They can be affectionate, and are drawn to people, but they are also restless and unpredictable.

Chloe challenged Maria's sense of communication more than Rocky or the donkeys did. Sometimes she wants to ride, sometimes she doesn't. Sometimes she wants to go forward, sometimes she wants to go backward.

Maria was frustrated working with Chloe at first, even angry sometimes. Her breakthrough came when she cleared her head and ego of the idea that Chloe was thwarting her. She learned to be patient, to be clear, to move her body in different ways. She called in friends who knew horses well, and they gave her tips on where to look, how to hold her body, how to give commands in a way the pony understood.

After a few weeks, Maria began exploring different ways to listen to Chloe and talk to her. She discovered that Chloe loved to have her tail groomed; the pony stood still for it and her lips quivered with pleasure. She also loved to have her mane brushed. Maria taught her to "kiss," to touch noses, and to "push" — to push balls and sticks around with her nose. She taught her to stand calmly while being cleaned or brushed.

Maria would take Chloe out of the pasture and walk her around the big old trees, helping the pony to pick some apples off a tree, to walk and turn together, to listen to each other. She opened her heart to Chloe, stood with her silently so Chloe could smell and sense her. Maria imagined riding her around the perimeter of the pasture. She pictured Chloe moving in sync with her, and it began to happen.

In various ways, Maria told Chloe that she cared for her, trusted her, that she was safe, a reliable source of food, grooming, attention, and stimulation.

Chloe was an even greater challenge for me than she was for Maria. When I would open the pasture gate, Chloe would come rushing toward me, grabbing at the hay, sometimes knocking me off-balance.

This was new to me — the donkeys and sheep never behaved in this way. And Chloe is big and strong; she could easily knock me or Maria over or step on our feet with her heavy hooves.

At age sixty-eight, I am happy to do the farm chores with my wife, but also a bit more cautious and defensive than I used to be. I swatted at Chloe with one hand, tapping her on the nose. In that one moment, I dropped all of the techniques and ideas I had so laboriously been collecting, trying, and studying for fifteen years.

Maria laughed. "Wait," she said. "Let's practice what you preach. Let's do with Chloe what you always do with the dogs. Pretend she is a dog coming near the road."

The lightbulb came on over my head. Still, for the next few days, I left the hay carrying to Maria while I watched what she did.

Maria carried the hay to the gate and opened it. Chloe came running up, whinnying in excitement. The horse turned and walked side by side with Maria to the hay feeder, accompanying her, sometimes running up happily ahead of her to wait at the feeder for the hay.

"Great, great," I said. "How did you do that?"

Of course, I had a good idea how. Maria imagined the scene she wanted to unfold, conjured up images of Chloe walking in just that way. She cleared her own head. She didn't go into the situation expecting conflict or discipline, or showing fear or hesitation, as I had. She was calm, confident, and clear. She showed these feelings to Chloe and focused on the horse's excitement, not her overeagerness.

Maria projected trust, looked toward the feeder, moved steadily toward it. And you could almost see Chloe catch on: Oh, I see, this is how we are going to do this. I will get the hay, it will be over there, and we will have a good time walking there together.

And from the pony's point of view, what could be better? Fresh hay in the cold, her human carrying it, the two of them walking to the feeder. Every day, the same time in the same way. Trust it, Maria said to Chloe, and so she did.

Chloe and the horses of Blue Star speak to Maria often. They tell her about myths and images, feelings and fears, joys and triumphs. They inform her art, nourish her spirit, show her the way to strength and patience and love.

I was more of a skeptic, at least initially. For months before Chloe came, I wondered why we were getting a horse. Did we need one? Could we handle another equine?

I can no longer imagine living without our horse. Chloe has strengthened our connection to the world of animals. She adds her voice to the chorus of horses who speak to all of us, who need all of us. Together, I believe people and animals can help heal the world, just as humans and animals built it together. That is what the horses are telling me, every single day.

In the spring of 2016, I watched Maria teach Chloe how to kiss. She held up a piece of carrot, said "kiss," and stuck her nose out in front of the pony. Chloe figured out in a flash that if she touched noses with Maria, she'd get a carrot or a cookie.

I wondered at the simplicity of this, this big creature happy to kiss a human on command for a cookie, or just for fun. Every morning of my life now, I get dressed and walk out to the pasture gate. Sometimes I have a carrot, sometimes not.

"Good morning," I say, "kiss!" And I stick my head over the gate. Chloe comes trotting up to me and kisses me on the nose. I have never found a better way to start my day.

11
FATE

My newest dog, Fate, is on the opposite end of the dog spectrum from Lucky, my first dog, the dog that began my journey with animals.

Lucky was a mongrel born in a box in a tenement basement in Providence, Rhode Island. Fate is a purebred border collie whose mother and father came from Wales and whose breeder is one of the most respected border collie breeders in America.

If Fate and Lucky are at opposite ends of the spectrum, so am I, decades away from that frightened little boy who got up early on a freezing-cold day to get his first dog.

Look what you started, Lucky, and look what Fate has wrought.

In the winter of 2015, we lost three animals we dearly loved: Simon, the donkey, who was rescued from an Upstate New York farm by the New York State Police; Lenore,

308

a loving Lab who kept love alive for me when I most needed it; and Frieda, the hell dog, who protected Maria every minute and kept every man except for me far away from her.

But this is the miraculous thing about dogs and other animals: unlike people, you can get another one. It sounds a bit cold, but it is the truth, a profound element in the human–animal bond. If you live with animals, you will know death, life, loss, love, and rebirth. We lost Lenore, Rocky, and Simon. Then there was Red.

My border collie, Red, had come from an amazing breeder named Karen Thompson, who lives in Virginia and owns and runs Thompson's Border Collies. Red is a wonderful dog, sweet and trustworthy, and a great worker. After some fruitless months looking online and talking to rescue groups, I emailed Karen and said Maria and I were looking for a dog.

She answered me immediately. As fate would have it, she said, there was a puppy suddenly available. She sent a photograph. The puppy had one black eye; the other was bright blue — the Merle gene. She looked like a pirate, full of herself, clearly a rascal and a handful.

Maria and I both at once said, yes, this is

our dog. A couple of weeks later, we drove down to Virginia to pick her up and bring her home.

Just as Chloe challenged Maria to test her work in communications, so Fate would challenge mine. I misunderstood Karen when I first spoke with her about the puppy. When she said "as fate would have it," I thought she meant the dog's name was Fate. It wasn't, but it is now.

Fate comes from a high-octane herding line. I have never had a more explosive or energetic dog. For the first four or five months of her life, she never stopped moving, not for a minute, from the second she woke up until she rushed into her crate and collapsed at night.

Even by border collie standards, she is a perpetually busy dog, looking for work, stalking the sheep, chasing Red around the room, tossing her toys into the chair, emptying the wastebaskets in the house and piling the contents up on the floor, hiding her toys and bones under sofas and in big holes dug in the yard.

Fate is the kind of dog you can wreck in a minute, she is so intense and aware. It is very easy to damage a young and high-speed border collie. They are so sensitive and obsessive that the wrong voice, the wrong movement, or vented anger, frustration, or confusion can alter the way they work for life. If you are too loud, or train too soon, or too late, or go too quickly or slowly, they can easily be ruined as herding dogs. We've all seen crazy border collies running around in circles, chasing balls and Frisbees all day as their humans desperately try to occupy them and tire them out. That was a lifestyle I wished to spare Fate.

Fate was too intense to train much at first. She was so full of instinct and drive that she could barely hear me, let alone respond. She tore off after the chickens, scouring the grass for chicken droppings. She went from room to room, looking in the wastebaskets for tissues to chew. Out in the pasture, near sheep, she could not even hear me when I shouted commands to her. When she heard the pasture gate open, she exploded forward, ignoring repeated commands to lie down and stay.

When she came into the house, she jumped up on me or any other person she saw. She leapt onto chairs, into laps. Sometimes she overshot my chair, bounced into walls, knocked over lamps.

Once she realized there was food on the counters, she would leap straight up into the air and try to snatch it on the fly.

I understood the value of crates. We had several in the house, and when she got too crazy, she went inside one of them. She also slept in them.

This dog, I immediately grasped, would either be one of the greatest dogs I ever had, or a rolling disaster. Border collies like this are often too much to handle for many people. When I saw how intense and frantic she was, I wondered if I could handle it. But I also saw the possibilities. Here was a living laboratory for my ideas and techniques about talking to animals. Training a young puppy so full of herself and with herding lines to work with stock is, to me, one of the greatest challenges in any line of work. When young border collies get near sheep, they are transformed. They enter a kind of trance, racing in frantic circles, tongues on the ground, panting and pausing to eat as many sheep droppings as they can find. They shake and tremble; they hear and see nothing beyond the sheep in front of them.

There are so many things they must learn: First, they must herd the sheep with a herder, not by themselves. Second, herding sheep is not a marathon race, a high-speed dog equivalent of a video game, which is what the inside of a border collie puppy's head often looks like, I believe. They must move the sheep but not run them, control them but not outpace them or panic them or harm them, all the while fighting belligerent rams, protective ewes, clueless lambs, heat, cold, ticks and bugs, barbed wire and groundhog holes, rocks and bushes and thorns. And of course, humans, who are often distant, distracted, confused, or inconsistent.

Many of us have seen those beautiful videos of superbly trained Irish and New Zealand dogs rushing up and down faraway hills, responding instantly to the whistles and hand signals of their humans. That's like comparing those diaper ads with cooing babies to the real lives of mothers and fathers with newborn infants. One is not recognizable to the other.

Training a dog like Fate is not simple, pretty, or easy. It is complicated, messy, unpredictable, and exhausting. It requires continuous repetition, patience almost beyond the ability of someone like me. It demands clear goals and a focused mind.

Taking care of sheep with herding dogs is something human beings have been doing since the beginning of recorded history. Despite its challenges, it does not ever feel alien. To me it feels like the most natural thing in the world.

Still, I never trained a dog as explosive or intense or as young as Fate. Red came to me almost fully trained. Rose and I worked together seamlessly. She was a fierce, hard-working, and intuitive farm dog, and we just figured it out together. As wonderful a dog as she was, she was not from the same pure working stock as Fate. Many if not most border collies bred in America are bred as show dogs or pets. They can be intense and energetic, but they are not the real thing.

Fate is the real thing, as was Red, who was bred in Northern Ireland. Border collie purists (or snobs, as I sometimes call them) are terrified that people like me will choose these show dogs and will not buy farms and sheep for them.

The purists were almost universally horrified when I wrote *A Dog Year* about my blundering efforts to figure out how to deal with Orson. And many were outraged when I ended up euthanizing him after he turned aggressive and bit several people, one in the neck. I do not regret putting Orson down, as much as I loved him. It was, for me, the beginning of a true understanding of the importance of talking to animals. But since his death, I have taken especially seriously the responsibility of living with border collies. Every one of them has lived on a farm, worked with sheep just about every day of their lives. They all had work to do, paths to run on, fields to explore, woods to race through, people to be with all day, cars to ride in, contact with all kinds of loving people.

Still, Fate represented the biggest training challenge I had yet faced. I could have found an expert to help me work with her, but I wanted to train her myself, for many reasons. For example, I never do well with formal instruction. I never had a single class in school or college (I dropped out) that I liked. I never felt close to a single teacher. I know that was my own issue, not a defect of the educational system. But still, it made me edgy.

For all of his flaws and contradictions, I am a disciple of Henry David Thoreau. I embrace the idea of making my own mistakes, learning from them, making my own decisions, standing in my own truth.

I wanted to communicate with this dog, teach her the ancient art of sheepherding, train her to live safely and lovingly with us, show her how to rest and be calm. And once in a while, to be comfortable doing nothing at all.

I wanted to be the teacher I never had,

and I wanted to begin a dialogue with her that would enrich both of our lives.

She was just eight weeks old when she first came into our house. She looked up at me sitting in my living room chair, came charging toward me, leapt into the air, and went sailing right over me, crashing into the wall, knocking over a lamp, bouncing to the floor.

Okay, I said to myself, *the journey begins.* I started practicing what I call "intuitive training," derived from my own instinct and intuition rather than from the books, videos, and dogmatic philosophies of others.

When I live with an animal, or seek to get to know one, or train one, or talk to one, I ask this first: What is its nature? What is the soul of this dog, this donkey, this ewe, this cat?

Like us, each animal is unique. Fate's nature is intensely reactive, energetic, curious. She is almost seething with impulse — to run, to chase, to chew, to move, to work. She is the most athletic and agile dog I have ever had. She can run for miles, leap five or six feet into the air. She flips objects high into the sky and chases them, prowls the house for things to find, move, and pick up. She is a problem solver and puzzle solver, as working dogs are.

Fate misses nothing, is curious about

everything. She has all the traits of the working dog — she eats donkey, sheep, and chicken droppings when aroused, is so excited around sheep that she can barely hear me or any human voice for that matter. She is drawn to people and energized by them, jumping up on every single person she meets, often repeatedly. At times, she is beyond reach or command.

My task was to understand this remarkable creature, to learn how to break through and communicate with her, and discover these instincts and channel them in a positive way. I needed to vary her activities so she would not become obsessed or addictive, to show her how to be calm when Maria and I worked or when we were in the house. I had to slow her down around sheep.

It takes a long time to teach a young border collie how to herd sheep properly and productively, so I decided to waste no time. Technically you are not supposed to start training border collies with sheep until they are at least six months old. My intuition told me that this would not be the case for Fate. I began working with her three days after she arrived, the minute I saw her discover the sheep and try to dig her way under the fence to get to them.

When I pulled her away, she threw herself

against the pasture fence and howled. Okay, I told Maria, I'm not going to do this to her for the next four months. She's going to work.

But first I introduced her to the crate, her quiet place, her resting place, her centering space. A place we could always put her when she got too excited, too aroused, too confused. Crates are dens to dogs, a natural and comforting habitat for them. People who think crating is cruel for dogs are being cruel to dogs. They are denying them the easiest, safest, most ancient way of being calm and feeling safe. There are no choices in crates, no confusion, no challenges and distractions. There a dog can feel calm, absorb the world around them, get to know him or herself.

Crate training is easy. With Fate, I put food inside the crate, and a few toys and treats. Every time she went inside, she found something to eat, something that made her feel this place was hers.

Food is not only nourishment to dogs — it is life. Anyplace a dog is fed is a home to the dog. Within a few hours, Fate was going to the crate on her own, happily resting after work or stimulation.

The crate was our first communication. I offered her a safe place to rest, eat, and sleep. I gave her something precious, and my taking her there associated me with good things, with safety and peacefulness.

Then we went out to the sheep. Fate just about blew her lid when she saw the sheep. Her eyes nearly popped out of her head; her body froze. She seemed to leave the earth behind and take off into space.

I decided to talk to her. I opened the pasture gate and let Red, the experienced dog, out first. I wanted to make sure the sheep didn't go after the puppy and mistake her for a raccoon or coyote, as they sometimes will. I didn't want her to get butted or chased or frightened. She was very young and very small.

I felt it was time for her to meet the sheep, and she needed me to take her there. So I let Red out of the gate and told him to get the sheep and put them in a corner of the pasture. He did what I asked.

Fate stood inside the gate staring, transfixed, then she shrieked in protest and raced back and forth, desperate to follow Red.

When the sheep were in their corner, I opened the gate and Fate shot out, toward Red and toward the sheep. When she got closer to them, I put her on a long lead and together we began marching toward the sheep. At first, Fate barked, shook, trembled, and lunged.

When she did, I stopped. Every few feet, I lay down on the ground. I was mimicking the behavior I wanted her to adopt. After the fourth time, she came over to me, fascinated, and lay down. I praised her effusively, with words, with touch, with body language, and she has consistently lain down on command ever since.

But we weren't there yet. I calmed myself and began speaking to her in a low voice. I imagined the two of us sitting quietly in the pasture, this whirling and lunging puppy now still and meditative. It took about fifteen minutes. I kept telling myself to be calm, be patient. Once or twice, I pulled on her lead in frustration, trying to get her to be still and lie down. Of course, this was a mistake, and it only made her more frantic. But this is the reality of training. They get it one day, and not the next. Circumstances change, the sheep react differently, it is hotter or colder, the trainer is confused or in a bad mood. Training doesn't just happen when you get it right the first time. It goes on and on, often throughout the life of the dog.

But I finally caught myself and truly let go of impatience. I just lay down and held the lead loosely and after a few more minutes of lunging, Fate lay down and stared at the sheep. They stared back. Red, behind us, kept everyone in place. The image I had in my head began to come to life in the real world.

It was a beautiful, warm, and breezy late spring day. The sheep stared curiously at Fate, then lowered their heads to graze. They had been herded by dogs almost every day of their lives. Fate was not nearly strong enough in eye or body to make them obey her. They were not impressed by the presence of this hyperkinetic puppy.

We sat still, soaking up the sun, feeling the soft breeze, being calm.

And that was the message I was bringing to Fate, the excitable puppy: it's okay to be calm, okay to do nothing, to be still, to watch, to learn.

And the message I received back, from her eyes, her body, her ears, was this: *It is nice to sit here with you, it is safe and good to be still, to watch the sheep, to test them with my eyes.*

It was an important bonding moment, setting the tone for that day and for all that was to come.

Recently, I started taking photographs of the "posted" signs that sprout every year just before hunting season, when many people decide whether to allow hunting on their property. The signs are "posted" to let the hunters know when they are welcome and when they are not. Some of the signs are yellow, some red, some handwritten or painted on wooden posts. The old ones often have vines and flowers growing around them; they are evocative and timeless. I did a photo series called "Post Art" and published the pictures on my blog.

Fate, a curious, even vigilant, puppy, watched me as I paused by the "posted" signs and took some pictures. It wasn't long before she was joining in the work. She would run ahead, find a sign by the side of the road, and stand beside it, waiting for me to come up with the camera. She never noticed the signs or stood by them before, but now she proved expert at finding them. Sometimes she posed, wanting to be part of the photo; sometimes she just touched the post with her nose and then sat down until I came up.

While Fate took on the task of spotting my signs, Red, my other border collie, would sit by my side while I considered the angles and took the photo. It is a wonderful gift for a photographer to have a dog — an active working dog, no less — who will sit faithfully and absolutely still while he or she stands by the side of a country road and takes pictures.

It is hard to relate what these dogs do for me and my work and do it justice. Fate and Red keep me company, watch over me, inspire me, guide me to the creative spark. They move the sheep around, too, and ride shotgun in the car. They remind me of all I've come to understand about communication with animals.

I understand the connective power of food.

I know the importance of getting in touch with my own emotions in order to understand theirs.

I have learned never to see animals as pitiable and abused creatures, but rather as my partners on the earth.

I know we cannot heal ourselves or the earth if we do not preserve and protect them.

I understand that it is not possible to love animals or find a place for them on the earth if we do not also learn to love and support the people who care for them, live with them, work with them.

I know that saving animals does not simply mean to save them from human cruelty and abuse; it means sacrificing, changing, and opening our minds to the idea that they are as important to us as cars, trucks, buses, highways, and condominium towers.

I see we need a new awakening, a new kind of movement to speak for the animals and to define and protect their most basic right: to remain in our world and in our everyday lives.

I know that work is not cruel for domesticated animals, but essential in order for them to survive and to give human beings the drive and motivation to take care of them.

Saving animals does not mean taking them away from people, forcing them out of work, and out of cities and suburbs and towns. It means just the opposite: it means finding work for them everywhere we can and as often as we can. If we brought animals like horses back into our lives as we have brought dogs to the center of our lives, then we would quickly come to know them and feel the same way about them, and no one would dare to ban or banish them or take them away from us in the name of protecting their rights.

Every animal banned or banished is another drop of blood for mother earth and for the idea of the great partnership between people and animals that has existed for all of human history. This partnership is being destroyed by a new idea of animal rights that does not protect the rights of animals, and by well-meaning human beings quick to exploit animals in the name of righteousness and noble feeling but slow to understand them or learn what it is they truly need.

Once again, we need a better and wiser understanding of animals than this, one that honors what Henry Beston wrote nearly a century ago. We cannot make good decisions about animals if we simply project our own fears, emotions, neuroses, and fantasies onto them.

They are not, in fact, our dependents. They are not our siblings or children. They are and have always been our partners in the joys and travails of the world, and they need now more than ever for us to understand them in truthful and accurate ways.

They have a right to be understood. We are morally obliged to look harder for ways to keep them in our world and in our everyday lives, rather than seek ways to remove them from the world.

We got Fate when she was two months old, in May 2015. Every day for seven months — sun, rain, snow — I took Fate out to work with the sheep.

Young border collies are easily aroused — they are impulsive, obsessive, and easily distracted. A border collie around sheep can be a wild animal, challenging the patience, clarity, and instincts of the human beings training them. But Fate was more than game and we went successfully through the usual training dramas and challenges. Getting her to pay attention to me. Getting her to lie down on command. Getting her to slow down. Moving in relationship to her. (Border collies move in relation to sheep and their human shepherd; it is a ballet, not a solo, and they all move a lot faster than I do.) Helping her learn to slow down, to give eye, to not be pushed too far or too fast, where she might be intimidated by the sheep.

I was determined to be patient, to reinforce her every good move, to wait for her to grow up a bit and be able to focus and concentrate more, to show more authority to the sheep, to get behind them and drive them, move them forward the way the dogs always do on television, the way Red and Rose always did.

But as the months passed, I became uneasy. Something was wrong. Fate was not evolving in the way herding dogs do, in the way my other dogs had done. She is a wonderful dog, loving, intelligent, fun. But the sheep just did not seem to be responding to her the way they always do to my border collies.

I thought about attachment theory, I thought about communicating. I began to ask myself some questions, rather than focus all of my intentions and expectations on her. I asked myself: What did I want from this dog? Was I expecting another Red, the perfect working dog in almost every way? Was I playing to the crowd, perhaps worrying more about the people watching me than the dog I was supposed to love and train?

I consulted with two friends who are herding trainers and asked them to come to the farm to backstop me. They said I was doing everything right, that the training was good. They both noticed that Fate seemed to be having fun but did not seem all that interested in the herding part. She gave strong eye, but she didn't seem to mean it the way Red and most border collies mean it.

See if that develops, they said, politely urging me to consider her interest in the nitty-gritty details of sheepherding.

With that in mind, I decided to take my own advice and talk to her, to see if I could get a sense from her what she wanted to do, what kind of dog she really was and wanted to be.

I took Fate out in the pasture one cold February morning. I didn't give the usual commands. Instead I released her — told her she was free — and she ran out into the pasture and began circling the sheep. I didn't walk her up to them, ask her to slow down, or tell her to "give eye."

When I watched her, I saw that she wasn't moving in; she wasn't developing authority. But she *was* having fun. She burst out running, circling and circling the sheep. Every time she got close to them, they either challenged her or blew her off. This would have made Red crazy, but it didn't bother Fate at all. Released of the burden of moving the sheep, she was having the time of her life, and the sheep were happily grazing, comfortable around this strange little creature who was practically dancing all around them.

I called her to me, and we sat down on the ground together. I closed my eyes, focused on her, imagined her herding the sheep in classic border collie fashion, then changed the image, and imagined her running, walking alongside me and Maria, being near the sheep but not really responsible for moving them.

Suddenly I felt a strong sense of release myself, free of this burden of training, the sense of pushing a boulder uphill, the feeling of straining and pushing this wonderful creature into being something she was not.

I thought of my father, who had pushed me so hard to be an athlete, calling me a sissy, undermining my confidence and character. I thought of Homer, the ungainly little dog who kept lagging behind on our walks.

I asked myself one of the best questions I could ask when talking to an animal: was I doing this for them, or for me?

I loved Fate very much and I was thrilled to have her and be with her. She was perfect for me and for Maria, a family dog who could go anywhere and do anything. She did not have a mean or aggressive bone in her body.

That day, I got the answer I had been seeking. It was simple, and I was sure it was coming from her. The message came to me clear as a bell:

I'm not Red. I don't want to be Red. That was it. That was all I heard, and it was more than enough.

I dropped the herding training that morning. Despite her lineage I gave up on the idea that she was going be a working dog.

She was going to be a happy dog, a family dog, a loved and appreciated dog. To this day, she still loves working with us. Every morning she rushes to the gate, runs to the sheep, greets them, circles around them, runs back and forth around the pasture, wears herself out, and has a blast.

So do we. My relationship to Fate has changed, just as my own relationship with my father might have changed if he had listened to me. If he had let me be who I was, and not who he wanted me to be.

We live in sweet harmony with Fate. There is nothing between us but love and connection. She comes when called, sits when asked, stays when commanded. She never runs off or goes near the road. "Wow," I've told Maria, "we have the dog we wanted." We always did, but I didn't even know it until I listened.

EPILOGUE:
WHAT THEY TOLD ME,
WHAT THEY TAUGHT ME

I began talking to animals more than half a century ago, the day when Lucky began one of the most spiritual and transformative periods of my life. I wish I could go back in time and talk to Lucky. I wish I could talk to the frightened but willful little boy who fought so hard to get him, and who learned so much from him.

Look what happened, I want to say. *Look what you started.*

On a fall day not long ago, I took my dogs, Fate and Red, out to the deep woods to reflect on the decades I've spent living in harmony with animals, to think about how far I'd come. The woods are our cathedral, our holy place. There is an old stone bench far out in the woods that was built many years ago, perhaps for the purposes of rest and contemplation.

I sat down on the bench, put my camera down, and watched the sunlight burst through the forest canopy. I listened to the dew falling off the trees, watched the dogs chase hopelessly after the nimble chipmunks, and heard a hawk's lonely cry.

At this point, the dogs know me; we understand each other. When I sit down like that, they check out every smell on the ground, then come over to sit down nearby. Sometimes they need to think about things, too. It is a beautiful place to sit, and I am grateful to have been introduced to it by Maria, who discovered the bench on one of her walks and brought me there. It is a sacred place.

Lucky

I thought of Lucky first, of course. Lucky told me that I was not what I feared I was, not what my father told me I was, not what the bullies screamed at me on the way home. He told me I had strength, and that there were good people in the world, that good things could happen. He made me see that there was love in the world for me and that I was a good person whose story was important.

Orson

Orson, my border collie, told me that I needed to change my life. *You are unhappy,* he said. *Your spirit is draining out of you, you are giving up on life. You do not belong where you are living, you are called to your adventure. It is never too late to change. It is time to leave the ordinary world behind, to set out on your hero journey.*

Julius and Stanley

My yellow Labs, Julius and Stanley, were the first dogs in my adult life who helped me become a writer. When I spent a year living alone with them in my upstate cabin — the beginning of my adventure — they were with me every moment. They sat with me through a difficult winter, were by my side when I wrote, sat with me outside when I read my Thomas Merton journals, walked with me in the woods as I learned how to live in nature. They seemed to tell me that it was natural for me to be a writer, that I could do it. In accepting my new life so graciously, they helped give me the strength to live. They marked this critical passage in my life, and left me when it was done. I will always recall the superstorms of that winter, all three of us curled up in bed, the snow piling up outside the windows, the wind shrieking through the chimney.

Elvis

Elvis, the Swiss steer, taught me to be wiser and more thoughtful about my life with animals. *We are not all pets,* he said. *Big and powerful and hungry animals like me are not meant to be companions for people. We have a different destiny.*

Winston

In many parts of the animal world, we are taught that it is never humane to help animals leave the world. Winston, my ten-year-old rooster, taught me a different lesson. Early in his life, his right leg was injured when he tried to defend his hens from a hawk attack. He limped for the rest of his days. He was a gentle, regal, but conscientious rooster, always looking out for his hens. When he fell ill at the end of his life, he told me he didn't want to die with the other chickens pecking at his eyes, as they do with sick brethren. He lay dying for two days, protected by the barn cats, who loved him. I heard his message. I took him out to the pasture, dug a hole in the ground, shot him, and, after saying a few words of respect and gratitude, laid him down. I could almost feel him thanking me. Mercy means many different things in the animal world.

Rose

Rose, my other border collie, told me that I could live on my farm, be in the country, do something I never imagined doing. She showed me how far I had come since Lucky, how much I needed her help, how she empowered me to survive those years alone, those awful blizzards, those belligerent rams. To hear the cries of the newborn lambs, to open my heart to the ancient glory of the shepherd, to walk our sheep through the woods and down the faraway paths. To move in harmony, as human beings have been doing with dogs and sheep since the beginning of recorded time.

Simon

Simon the donkey opened me up, as much or more than any animal since Lucky. For a time, I wondered sometimes if he might be Lucky reincarnated. Simon told me I could put my hand in his mouth and put ointment on his painful and swollen gums. He told me he trusted me to put drops directly into his badly infected eyes. He told me he trusted me to pour balm on his sores and open wounds. In doing so, he led me to new experiences, to agreeing to take Red, even though I had never seen him. To walk in the woods with him, as the ancients did with their donkeys. In healing him, I healed myself. He also drew me more deeply into the wonderful world of equines, which led me to stop and take a picture of the blind pony Rocky, which led me to Maria and to buy the farm Rocky loved. *Open up,* Simon told me over and over again, in many different ways. *Open up.* And so I did.

Ma

When Ma, the ewe, was struggling while giving birth, I wondered if it would be most merciful to put her down. As she lay gasping for breath in labor, I heard her message very clearly in my mind, and felt it in my heart and soul: *I have two babies in my womb, please let me live.* She showed me how to pull one out, and then the other. They both lived, and Ma climbed to her feet. Exhausted and bleeding from the inside, she licked them both clean and lived long enough to give them the milk they needed to start their own lives.

Frieda

Frieda, the man-repeller, told me that she would always protect Maria, and she would trust me when Maria trusted me. She showed me the way to Maria's heart by learning to love me and, eventually, trust me. She opened my heart to love, and its possibility. She gave me the courage and strength to persevere with an animal that seemed impossible, even frightening to work with. She made me understand how important love was to me, and how patient I needed to be in order to have it. And she showed Maria that I was worthy of love and trust when both of them had struggled so long to find it. She brought me to love. Maria told me later that she trusted me when she saw how I showed Frieda how to trust me. And I did it by imagining what I wanted, and it came to be.

Rebecca

I met Rebecca, a beautiful draft horse rescue from an auction house, in a New York carriage horse stable. She looked at me in a particular way, and when I left New York to come back to my farm, Rebecca's voice came with me. *We are content here,* she said. *We are living our destiny, to work with people and be active and engaged and have a purpose. If we leave,* she told me in the middle of one dark night, *we will take the wind and water and fire with us; we will take the last drop of magic from the big city. Do not let them take us away and hide us from view again. We are the partners of people; we share the joys and travails of life. Tell our story, speak for us.*

Red

Red came to show me the great heart of animals, their generosity of spirit, their mystical ability to enter the lives of human beings and walk with them through life. Red is a therapy dog; he lifts the spirits of troubled people. He is a working dog; he helps us live on our farm. He is a companion dog; he accompanies me on my journey through life. He is a spirit dog, here to mark a passage of my life. He is a muse; he unlocks the spirits, chases away the demons, and helps me to write.

Rocky

Rocky, my blind Appaloosa pony, had lived alone on our farm for many years. He was sick and tired, and he told me and Maria when it was time for him to go. He could not face another winter. He could not survive the determination of the donkeys to drive him away because he was sick and infirm. He taught me that the quality of mercy is not strained. He begged us to let him go and join his ancestors rather than keep him alive and struggle so that we could feel good about ourselves. He reminded me once again that we don't have to guess or argue about what it is animals need and want.

If we listen, they can tell us.

ACKNOWLEDGMENTS

I want to thank Maria, and the very many people who have shared my life with animals, taught me how to live with them, came to my aid when I needed them. There are too many to list without leaving out others or inadvertently hurting some.

But some have to be named.

I wish to thank Mr. Wisnewski, the janitor at the Summit Avenue Elementary School; he stood up for me and started me on my journey with animals.

I wish to thank the vets: the Granville Veterinary Service–Large Animals; Tom Wolski and Suzanne Fariello and the Cambridge Veterinary Service; Ken Norman, our farrier; Jim McCrae, our shearer; Peter Hanks, who brought me Elvis; Karen Thompson, who brought me Red and Fate; Gretchen Pinkel, who gave us Lenore.

I want to thank the many farmers who befriended me, took pity on me, taught me what I needed to know, sold me good hay, taught me about grass and the seasons.

I want to especially thank Ed and Carol Gulley, our friends and hardworking dairy farmers, for their friendship, inspiration, and support. And Kim and Jack Macmillan, our neighbors and friends, who have rushed to our aid too many times to count.

I thank my close friend Scott Carrino, whose friendship has sustained and nourished me in my new life. He is a wonderful friend.

I thank Bruce Tracy and Brian McLendon, both formerly of Random House, for publishing my first animal books and supporting me, and Richard Abate.

I am especially grateful to Rosemary Ahern and Christopher Schelling for representing me, standing by me, editing me. And Peter Borland of Simon & Schuster for believing in me and this book.

I thank all of the wonderful people who have read my blog, supported me and it, and encouraged me to keep writing books when I wasn't sure I could.

ABOUT THE AUTHOR

Jon Katz has written twenty-seven books including works of fiction and nonfiction, both for adults and children — among them *The Second Chance Dog: A Love Story* and *The Dogs of Bedlam Farm*. He has written for *The New York Times*, *Slate*, *Rolling Stone*, and *Wired*. He lives on Bedlam Farm in upstate New York with the artist Maria Wulf; his dogs, barn cats, donkeys, sheep, and chickens. His last book was *Saving Simon: How A Rescue Donkey Taught Me the Meaning of Compassion*. He is also a photographer. Learn more at BedlamFarm.com.